Seawater in Concrete Mix

Seawater in Concrete Mix

Hidenori Hamada
Nobuaki Otsuki
Takahiro Nishida

CRC Press
Taylor & Francis Group
Boca Raton   London   New York

CRC Press is an imprint of the
Taylor & Francis Group, an **informa** business

First edition published 2022
by CRC Press
6000 Broken Sound Parkway NW, Suite 300, Boca Raton, FL 33487-2742

and by CRC Press
2 Park Square, Milton Park, Abingdon, Oxon, OX14 4RN

© 2022 Hidenori Hamada, Nobuaki Otsuki, and Takahiro Nishida

CRC Press is an imprint of Taylor & Francis Group, LLC

Library of Congress Cataloging-in-Publication Data

Names: Hamada, Hidenori, author. I Otsuki, Nobuaki, 1951- author. I Nishida, Takahiro, author.
Title: Seawater in concrete mix / Hidenori Hamada, Nobuaki Otsuki, Takahiro Nishida.
Description: First edition. I Boca Raton : CRC Press, 2022. I Includes bibliographical references and index.
Identifiers: LCCN 2021020235 (print) I LCCN 2021020236 (ebook) I ISBN 9781032046679 (hbk) I ISBN 9781032046693 (pbk) I ISBN 9781003194163 (ebk)
Subjects: LCSH: Concrete--Additives. I Concrete--Chemistry. I Hydration. I Seawater.
Classification: LCC TA441 .H36 2022 (print) I LCC TA441 (ebook) I DDC 620.1/36--dc23
LC record available at https://lccn.loc.gov/2021020235
LC ebook record available at https://lccn.loc.gov/2021020236

ISBN: 978-1-032-04667-9 (hbk)
ISBN: 978-1-032-04669-3 (pbk)
ISBN: 978-1-003-19416-3 (ebk)

DOI: 10.1201/9781003194163

Typeset in Sabon
by Deanta Global Publishing Services, Chennai, India

Contents

PART II
Actual Constructions   73

4 Japanese Experience with Seawater Concrete   75

Preface

Concrete has come into widespread use around the world for the construction of infrastructure, just like iron and steel. It is normally a mixture of cement, aggregates (fine and coarse), water, and various admixtures. Currently, only freshwater (tap water or spring water) is used in concrete mixes, while seawater has been avoided because of its salinity.

This book intends to show that concrete could be made with seawater instead of freshwater. Taking the positive perspective that this could be good for concrete production, both in terms of the concrete itself and from a sustainability point of view, the authors believe that 'where there is a will, there is a way.' We strongly believe that some structures and some situations are suitable for the use of seawater for concrete production. One example is infrastructure in a marine environment, where concrete structures are affected by seawater splashing.

Seawater in Concrete Mix is not in immediate demand, but we have no doubt that it will be needed sometime soon. The reason for this is a natural question for many people, even concrete engineers. It is clear that in the near future most people in the world will be suffering from water scarcity, with some underwater stress and many in water crisis. At the same time, seawater will still be available in abundance. Effective use of seawater in concrete production could be a key part of solving this crisis. In our view, based on experience in concrete engineering, there are a few disadvantages in using seawater as mixing water/curing water in concrete, but at the same time there are also advantages.

In the past several decades, it has been widely acknowledged in engineering that better materials should be used if they could be obtained easily. However, since early 20th century, especially in coastal areas or on isolated islands, seawater has been used in the construction of infrastructure, and in some cases construction itself has been done under marine conditions. Even now, some structures are still built using seawater in the concrete mix, particularly in the coastal areas of many countries. In a sense, the 20th century was a delightful era for engineers, who had an abundance of good materials available to them. However, in the 21st century, shortages of high-quality raw materials are beginning to be felt around the globe. The idea of the 3Rs (Reduce, Reuse, Recycle) has been promoted, meaning that all materials should be put to full use using all possible techniques.

One of the authors (N. Otsuki) started his research on seawater concrete in 1975. We found that seawater could be used in concrete production for marine environments, especially with ground granulated blast furnace slag (GGBFS), fly ash (FA), and various other mineral admixtures. Indeed, the most important technique in seawater concrete production is the usage of mineral admixtures. The recent trend toward performance-based design has been a great encouragement for us.

Without for the moment suggesting that seawater should be used for all concrete, our recommendation now is limited to the use of seawater in areas that are short of freshwater and for structures in marine environments, such as the following cases:

1) On remote islands
2) In seawater (or a nearby environment), including tidal and submerged zones
3) In coastal desert areas (very dry)
4) In arctic areas (very dry)

This book takes a positive position regarding the use of seawater in concrete. We expect a degree of skepticism and welcome for it as it demonstrates an interest in the topic. It is our hope that this book will generate interest in the use of seawater concrete.

Contributing Authors

This book is written not only by three authors but in collaboration with the following contributing authors. Without them, this book could not have been completed.

Nobufumi Takeda (Hiroshima Institute of Technology, Hiroshima, Japan)

He is Professor at Hiroshima Institute of Technology, Japan, since 2017. He has worked as Senior Research Engineer, Senior Research Fellow, at Technical Research Institute at Obayashi Corporation, Tokyo, Japan, and Professor at Hiroshima Institute of Technology, Hiroshima, Japan. His research topics are durability design of concrete structures, maintenance technology of existing concrete structures, utilization of industrial by-product and construction method of infrastructure.

Contributing chapters: 2, 3, 4 and 5.

Tsuyoshi Saito (Niigata University, Niigata, Japan)

He is Associate Professor at Niigata University, Japan, since 2012. He was Assistant Professor at Tokyo Institute of Technology, Japan. His research topics are cement chemistry and design of sustainable cementitious materials, based on chemical analysis.

Contributing chapter: 2.

Yoshikazu Akira (Kagoshima University, Kagoshima, Japan)

He is Associate Professor at Kagoshima University, Japan, since 2015. He was Researcher at Port and Airport Research Institute, Yokosuka, Japan, and Senior Researcher in Toyo Construction CO., Ltd., Japan. His research topics are corrosion and corrosion protection of concrete structures and steel structures, development of maintenance technologies of infra-structures, and development of high durable concrete, such as Shirasu concrete.

Contributing chapter: 4.

Takayuki Fukunaga (Kyushu University, Fukuoka, Japan)

He is Assistant Professor of Kyushu University, Japan, since 2020. He was Research Engineer at the Cement and Additives Research Laboratory at Denkikagaku Kogyo kabushikigaisya Nigata, Japan, and Technical staff at the National Institute of Technology Kagoshima College, Kagoshima, Japan. His research topics are cement chemistry of admixtures such as pozzolanic materials (especially Shirasu), durability of concrete using these materials and development of environmentally friendly construction materials using industrial wastes.

Contributing chapter: 4.

Takahiko Amino (Toa Corporation, Tokyo, Japan)

He is Leader of Materials and Renewal Group, Research and Development Center, TOA CORPORATON, Japan, since 2014. He has served in various roles such as Design Engineer of Design Department and Research Engineer, Senior Research Engineer, Leader of Materials and Renewal Group of TOA CORPORATION. His research topics are durability and maintenance of RC structures under marine conditions, life extension technology of severely deteriorated RC structures, and advanced material and construction techniques development.

Contributing chapter: 5.

Miren Etxeberria Larranaga (Universitat Politecnica de Catalunya, Barcelona TECH, Campus Nord UPC, Spain)

She is Associate Professor in Construction Materials and Sustainable Materials and Technology of Department of Civil and Environmental Engineering at UPC, Barcelona TECH. Her research focuses on recycling demolition, industrial or maritime waste, and using available and abundant natural resources (seawater and volcanic ash) for construction materials production. She has published more than 45 papers in JCR publications and international congresses. She has been responsible for national and international projects. In addition, she is involved in 5 international scientific committees.

Contributing chapter: 5.

Terminology

Seawater: water from sea or ocean. On average, the seawater in the world's oceans has a salinity of about 3.5% with an average density at the surface of 1.025 kg/L, making it denser than freshwater (1.0 kg/L at 4°C or 39°F).

Freshwater: any natural water excluding seawater and brackish water. Freshwater includes water, for example, in ice sheets, ice caps, glaciers, icebergs, bogs, ponds, lakes, rivers, streams, and groundwater (water present under the ground surface). Especially in this book, freshwater includes potable water (or drinking water).

Brackish water: water having more salinity than freshwater, but not as much as seawater. It may result from the mixing of seawater with freshwater, as in estuaries.

Water scarcity: lack of capacity to meet water demand. It is affecting every continent and was listed in 2019 by the World Economic Forum as one of the greatest global risks in terms of potential impact over the next decade.

Water stress: a shortage of water for the inhabitants of a country or area.

Water crisis: a health crisis resulting from lack of access to safe water and sanitation, which normally contribute to improved health and help prevent the spread of infectious diseases.

Concrete (cement concrete): a composite material consisting of fine and coarse aggregates bonded together with a cement paste that hardens over time.

Performance-based design: an approach to the design of any structure. A structure built according to performance-based design is required to meet certain measurable or predictable performance requirements, such as load-carrying capacity or cracking state, without a specific prescribed method by which to attain the requirements. This approach provides the freedom to develop tools and methods to evaluate the entire life cycle of the structure.

Defect: an identifiable unwanted condition that was not part of the original intent of design.

Deterioration: a defect that has occurred over a period of time.

Durability performance: one among several performance measures, including safety performance, serviceability performance, restorability performance, which a concrete structure must maintain for the designed service life. Durability is an inherent performance characteristic of concrete; however, it is affected by environmental conditions.

Service life: the term for which a (concrete) structure is designed to be in use in the specified condition.

Life-cycle management: the total management of a structure over all stages of its lifetime (planning, design, construction, maintenance, and demolition) by the most effective and most reasonable method.

Blast furnace slag (BFS): more correctly ground granulated blast furnace slag (GGBFS), a by-product of iron or steel making. GGBFS is obtained by quenching molten iron slag from a blast furnace in water or steam to produce a glassy, granular product that is then dried and ground into a fine powder. GGBFS is highly cementitious and improves the strength and durability of concrete.

Fly ash (FA): a by-product of coal combustion in power plants. A molten mineral residue is formed as the pulverized coal is ignited and this cools and hardens to form ash as the boiler tubes extract heat. The coarse ash particles fall to the bottom of the combustion chamber, while the fine ash particles remain suspended in flue gas. These finer particles, removed by precipitators, are called fly ash (FA).

Shirasu: a kind of pyroclastic flow deposit found in southern Kyushu, Japan. The original Japanese name *shirasu* is a general term referring to pyroclastic flow deposits from volcanic eruptions. It is from an old dialect of Kagoshima district, southern Kyushu island, in the far west of Japan.

Microcell corrosion: a microscopic cell formed on a continuous piece of metal consisting of an anode and a cathode immediately next to each other. This creates the electrochemical condition that makes corrosion possible. Corrosion microcells are formed due to impurities, environmental conditions, and other factors.

Macrocell corrosion: basically, a macrocell circuit (i.e., corrosion circuit) that occurs in steel objects enclosed in soil or a solidified concrete structure where the anode and the cathode are separated from each other. Macrocell corrosion occurs due to the presence of impurities, environmental conditions and other factors.

MEASUREMENT METHODS DESCRIBED IN THIS BOOK

Corrosion potential (mix potential)

Knowing the potential for a metal to corrode is very useful in corrosion studies. Fortunately, it can be readily measured in the laboratory or under field conditions. The corrosion potential is measured by determining the voltage difference between a metal immersed in a corrosive electrolyte and an appropriate reference electrode [1].

Linear polarization

Linear polarization (LP) is simple in principle, although the underlying theory and its use in practice are complex. The measurement of LP relies on the relationship between the half-cell potential of a piece of corroding steel and an external current applied to it. LP is usually measured in aqueous solutions on small, uniformly corroding specimens [2].

Polarization resistance

The polarization resistance, R_p, of a steel reinforcement embedded in concrete is defined as the ratio between applied voltage (shift in potential from E_{corr}) and the current increment, when the metal is slightly polarized (about 20–50 mV) from its free corrosion potential, E_{corr}. It can also be defined as the slope of the potential-current polarization curve at the corrosion potential, E_{corr} [3].

SPECIAL WORD DEFINED IN THIS BOOK

Seawater concrete: a concrete produced using seawater as mixing water

Freshwater concrete: a concrete produced using freshwater as mixing water

BIBLIOGRAPHY

1. NACE International. 2005. Corrosion basics, corrosion potential. *Materials Performance (MP)*, Vol.44. No.8:66.
2. John P. Broomfield, Kevin Davies, Karel Hladky. 2002. Permanent corrosion monitoring in new and existing reinforced concrete structures. Cement & Concrete Composites, Vol.24:66–71.
3. C. Andrade, C. Alonso. 2004. Test methods for on-site corrosion rate measurement of steel reinforcement in concrete by means of the polarization resistance method. Materiaux et Constructions, Vol.37:623–643

Acknowledgments

This book is basically from JCI Technical committee report. The committee chairperson was Nobuaki Otsuki, and vice chairperson was Hidenori Hamada. The book includes many parts of the committee report, and also new knowledge other than the committee activity. The authors express their sincere thanks to JCI, Japan Concrete Institute. Also, our thanks go to all committee members.

BIBLIOGRAPHY

1. Japan Concrete Institute [JCI]. 2015. *JCI Technical Committee Report on the Use of Seawater in Concrete*, ISBN 978-4-86384-067-6-C3050, September 2015.

Authors

Hidenori Hamada is Professor at Kyushu University, Japan, since 2009. He has previously worked as Senior Research Engineer, Head of Materials Division, and a member of Port and Airport Research Institute, Yokosuka, Japan. His research topics include durability of RC and PC structures under marine conditions, life extension technology of severely deteriorated RC and PC structures, and eco-friendly concrete material development such as seawater utilization in concrete production.

Nobuaki Otsuki is Professor Emeritus of Tokyo Institute of Technology, Japan. He has worked as Senior Research Engineer and Head of Materials Division at Port and Airport Research Institute, Yokosuka, Japan. His research interests include the durability of RC and PC structures under marine conditions, life extension technology of severely deteriorated RC and PC structures, and eco-friendly concrete material development such as seawater utilization in concrete production.

Takahiro Nishida is Senior Researcher of the Japanese National Institute of Maritime, Port and Aviation Technology, and Port and Airport Research Institute, Japan. He has worked for several research institutes, including port and airport, electric power industry, and highways. He has taught construction material at Tokyo Institute of Technology and Kyoto University. His research interests include the durability and maintenance of civil engineering structures.

Chapter 1

Introduction

1.1 SEAWATER CONCRETE

Seawater concrete is a concrete produced using seawater in the mix.

Why consider using seawater in concrete production? Currently, only freshwater is used to produce concrete in most countries of the world. However, it is certain that by the mid-21st century, a large number of people in many countries will be facing the realities of water scarcity, water stress, and water crisis, driven by rapid modernization in many developing countries and an exploding world population.

Most construction standards in countries around the world prohibit the use of seawater in concrete production, both for reinforced concrete (RC) and prestressed concrete (PC). However, it is a fact that many existing structures have been built with seawater concrete over the past five decades and more. Some examples of such structures in Japan are discussed later in this book: the Ukushima Island lighthouse in Nagasaki prefecture, the port breakwater constructed with pre-packed concrete (pre-placed aggregate concrete) at Tajiri Port, Tottori prefecture, and so on. Also, a trial of the application of seawater concrete for port concrete block, in Spain, Barcelona, is introduced.

1.2 THE CASE FOR SEAWATER CONCRETE

Seawater concrete is seawater mixed concrete. As already mentioned, seawater concrete is a type of concrete in which seawater is used as an alternative to freshwater for mixing the concrete. Using seawater inherently incorporates chloride ions in the finished concrete, which can be an advantage but also a disadvantage.

Sustainability is a major issue facing humankind in the 21st century and the non-sustainable depletion of resources is the main problem. Several billion tons of freshwater are consumed annually for the mixing, curing, and washing of concrete. Meanwhile, the abundant seawater available on the globe is presently not permitted to be used for these purposes, despite its easy accessibility in large quantities. Clearly, the active use of seawater by the concrete industry could contribute to the sustainability of freshwater resources, so seawater concrete could play a very important role now and in the future.

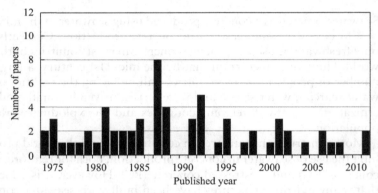

(a) Number of papers on seawater concrete published between 1974 and 2011 (T. Nishida et al.)

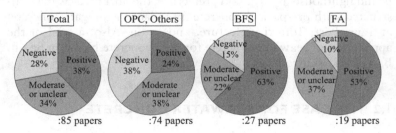

(b) Positive and negative on seawater concrete for various cement types

Figure 1.1 Literature survey on seawater concrete published between 1974 and 2011 (Nishida et al.).

1.3 IMPORTANCE OF MINERAL ADMIXTURES

Here we present the results of a literature survey of research into seawater concrete [1]. The literature on seawater concrete in the period between 1974 and 2011 consists of 68 articles, 32 in English and 36 in Japanese (Figure 1.1). Overall, in this body of work, seawater concrete containing blast furnace slag (BFS) or fly ash (FA) is evaluated as better compared to the concrete without BFS/FA, which is evaluated as not good.

We categorize the conclusions of these articles as 'positive,' 'negative,' or intermediate or unclear, as shown in Figure 1.1. Papers in which ordinary Portland cement (OPC) was used have a 'negative' conclusion in nearly 40% of cases. In contrast, where BFS or FA cement was used, more than 50% of conclusions are 'positive.' These mixed results make clear that the use of seawater as a mixing agent in concrete has its pros and cons. Nonetheless, many favor its use with respect to BFS or FA cement. Therefore, it is important to consider how to use mineral admixtures most effectively in seawater concrete.

BIBLIOGRAPHY

1. Takahiro Nishida, Nobuaki Otsuki, Hiroki Ohara and Zoulkanel Moussa Garba-Say. 2013. Some considerations for the applicability of seawater as mixing water in concrete. Proceedings of the Third International Conference on Sustainable Construction Materials and Technologies (SCMT3): CD-ROM e056.

Part I

Research and Technology

Chapter 2

Engineering Properties of Seawater Concrete (with OPC and BFS/FA)

2.1 HYDRATION

In this section, we report on the influences of seawater and mineral admixtures (BFS and FA) on the hydration of cement. The major points made below are: (1) the reaction properties of C_3A and C_4AF are significantly influenced by seawater, in contrast with the reaction of C_3S and C_2S; (2) in B20 (BFS is 20% in total cement) and B40 (40%) concretes mixed using seawater, the reaction ratios are especially high at almost twice that with distilled water; and (3) in cement mixed using seawater, there is no formation of mono-sulfate and ettringite by the age of 91 days, but a lot of Friedel's salt is formed.

2.1.1 Mechanism of Hydration Reaction of OPC in Seawater Concrete

(1) Reaction and Reaction Ratio of Various Cement Clinkers (C_3S, C_2S, C_3A, C_4AF)

The reactions and reaction ratios of the cement clinkers C_3S, C_2S, C_3A, and C_4AF in OPC have been reported by Saito et al. [1]. Their results are summarized below. In outline, the reactions of C_3A and C_4AF are significantly influenced by seawater at the initial stage of curing, while the influence on C_3S and C_2S reactions is much less.

Figure 2.1 shows the reaction ratio over time of each clinker in specimens N made with OPC only using distilled water knead-ing mixed (DW) and seawater kneading mixed (SW). Focusing on the reaction ratios of C_3S and C_2S (silicate phase), at the age of 3 days, representing the initial stage of the reaction, the DW and SW

DOI: 10.1201/9781003194163-2

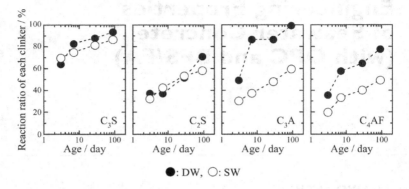

Figure 2.1 Reaction rate of each clinker in OPC paste specimens.

mixed specimens are similar, or the reaction rate of DW is higher than SW in some binders. The reaction ratios at 7 and 28 days continue to be quite similar. However, at the more advanced age of 91 days, the SW reaction ratio is lower than that of DW.

Next, looking at the difference in reaction ratios of the gap phases C_3A and C_4AF with DW and SW mixing, the reaction properties are found to differ from the silicate phase, with reaction ratios of each gap phase much lower for SW than DW from the initial stage of the reaction at 3 days. Moreover, with increasing age, the reaction is found to largely stagnate with SW, until at the age of 91 days a large difference of about 35% is observed between the SW and DW values for each gap phase.

These results demonstrate that the reaction properties of C_3A and C_4AF are significantly influenced at the initial stage of the reaction, compared with the reactions of the silicate phases C_3S and C_2S. It is considered that a dense gel hydrate containing SO_4^{2-} is produced in the C_3A phase.

(2) Hydration Products

In outline, the most distinctive difference in the hydration reaction in specimens N between SW and DW mixing is the type and amount of various aluminate hydrates produced in the gap phase. With SW mixing, the predominant hydration product is Friedel's salt, and almost no ettringite or monosulfate is produced by the age of 91 days. Looking at the changes over time in the amounts of the aluminate hydrates (ettringite (Ett), monosulfate (Ms), hydrogarnet (HG), hydrotalcite (HT), C_4AH_{13}, and Friedel's salt), the total at the age of 91 days for DW is about 17%, compared with only about 11% for SW. This result is consistent with the measured reaction ratios of C_3A and C_4AF, namely the gap phase reaction rate for SW is two-thirds that for DW.

Focusing on the phase constitution of the aluminate ferrite hydrates, compared with DW, the amount of hydrogarnet is slightly lower, but C_4AH_{13} is greatly reduced in the case of SW. On the other hand, looking at changes in the AFt and AFm phases, whereas ettringite is converted over time to monosulfate with DW, in the case of SW the formation of ettringite is observed at the age of 7 days, but it disappears thereafter and conversion to monosulfate does not occur. Further, there is almost no ettringite or monosulfate at the age of 91 days with SW, and Friedel's salt is predominant.

The formation of Friedel's salt, monosulfate, and ettringite has been investigated by Xu et al. [2]. XRD analysis of samples with the addition of 1.5 mass% NaCl as binder indicates that ettringite is not formed but Friedel's salt is. Monosulfate and ettringite are not formed in Saito's study, and this is consistent considering that the amount of NaCl added is approximately equivalent to 1.47 mass% of the binder. Goni has reported that dissolution of ettringite occurs if the curing material contains a high concentration of alkaline aqueous salt solutions. In the process of monosulfate ($C_3A \cdot CaSO_4 \cdot 12H_2O$) changing to Friedel's salt ($C_3A \cdot CaCl_2 \cdot 10H_2O$). SO_4^{2-} is released into the liquid phase. The SO_4^{2-} and OH^- maintain a charge balance between the layers of the monosulfate structure. The SO_4^{2-} removed from the pore solution is replaced with Cl^-. It is reported that, based on the results of pore solution composition analysis and XRD, the OH^- and SO_4^{2-} concentrations in the pore solution increase [3]. The same author, M. Kikuchi et al., also analyzes the pore solution ion composition in the cured product in a previous report, with 0.5 mol/L of NaCl aqueous solution used for water mixing. It is shown that the OH^- concentration increases. Moreover, the concentration of SO_4^{2-} is as much as 4–7 times higher compared to the use of distilled water [4]. Jones et al. have also reported similarly; the initial addition of NaCl leads to an increase in OH^- and SO_4^{2-} concentrations in the liquid phase [5].

In the study by Saito [1], monosulfate and ettringite are not produced with SW mixing because, in the presence of Na^+ and Cl^- from the seawater, the SO_4^{2-} readily leaves the ferritic aluminate hydrate of the solid phase. The SO_4^{2-} is released into the liquid phase by the action of Na^+ and Cl^- rather than ettringite. The Friedel's salt in the AFm phase is consumed in the formation of a dense gel hydrate C_3A with the emission of SO_4^{2-}. From the SO_4^{2-} concentration in gel-form hydrates, Friedel's salt, which is an AFm phase, tends to be formed.

2.1.2 Reaction of BFS and C₃S, C₂S, C₃A, C₄AF in Seawater Concrete

Figure 2.2 show the reaction ratios of C_3S, C_2S, C_3A, and C_4AF for various BFS and FA cements. B20, B40, and B70 are abbreviations for BFS cements with BFS replacement ratios of 20%, 40%, and 70%, respectively [1]. FA is the abbreviation for fly ash cement. The 7-day reaction ratios of these minerals with seawater are just a little larger than those with distilled water. However, at 28 days and 91 days, the reaction ratios in the seawater case are less than in the distilled water case.

In Figure 2.3(a), the reaction ratios of BFS are shown for B20, B40, and B70. The reaction ratios of BFS with seawater are much larger than with distilled water, being almost double those with distilled water in the cases of B20 and B40. The effects of BFS on promoting reactions are also reported in [6, 7], relating to the compressive strength and durability of concrete.

The amount of aluminate ferrite generated during hydration is about 30% higher than with OPC. This is believed due to the formation of aluminate ferrite in the hydration reaction of BFS.

2.1.3 Reaction of FA and C₃S, C₂S, C₃A, C₄AF in Seawater Concrete

Figure 2.3(b) shows the reaction ratios of FA cement over time (FA20, with a fly ash replacement ratio of 20%). The reaction ratio with seawater is a little larger than that with distilled water. The time-dependent changes of phase constituents are similar to those in the case of BFS, with aluminate ferrite hydration predominant.

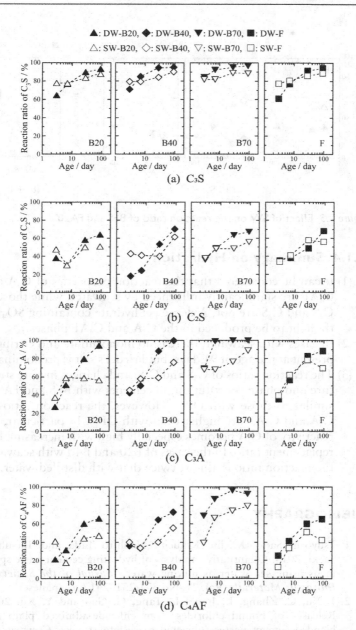

▲: DW-B20, ◆: DW-B40, ▼: DW-B70, ■: DW-F
△: SW-B20, ◇: SW-B40, ▽: SW-B70, □: SW-F

(a) C₃S

(b) C₂S

(c) C₃A

(d) C₄AF

Figure 2.2 Reaction ratio of C₃S, C₂S, C₃A and C₄AF in paste specimens containing BFS admixture.

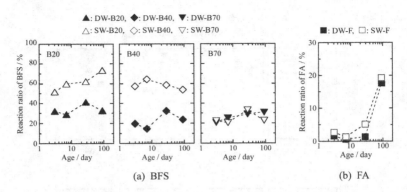

(a) BFS   (b) FA

Figure 2.3 Effect of SW on the reaction ratio of BFS and FA20.

2.1.4 Summary on Hydration

(1) It can be considered that the reaction properties of C_3A and C_4AF are significantly influenced by seawater, while those of C_3S and C_2S are not. A dense gel hydrate containing SO_4^{2-} is thought to be produced in the C_3A and C_4AF phases.

(2) In cases with seawater, there is no monosulfate and ettringite present at the age of 91 days, but Friedel's salt is predominant.

(3) The reaction ratios of C_3S and C_2S, and changes in pore structure and phase constitution, in cements with BFS and FA are similar to those with OPC. However, the reaction ratios of C_3A and C_4AF are higher than with OPC. In particular, the reactivity of C_3A becomes especially high with increasing BFS replacement ratio. In the cases of B20 and B40 with seawater, the reaction ratio is almost twice that with distilled water.

BIBLIOGRAPHY

1. Tsuyoshi Saito, Michio Kikuchi, Naohiro Tada and Tatsuhiko Saeki. 2014. Quantitative analysis of hydrating cement paste specimens using sea water as mixing water. *Journal of the Society of Inorganic Materials*, Vol. 21, No.371:231–241 (in Japanese).
2. J. Xu, C. Zhang, L. Jiang, L. Tang, G. Gao and Y. Xu. 2013. Releases of bound chlorides from chloride-admixed plain and blended cement pastes subjected to sulfate attacks. *Construction and Building Materials*, Vol. 45:53–59.

3. S. Goni, A. Guerrero and M.S. Hernandez. Spanish LLW and MLW disposal. 2001. Durability of cemented materials in (Na,K)Cl simulated radioactive liquid waste, Waste Management, 21:69–77.
4. Michio Kikuchi, Takayoshi Kanazawa, Tsuyoshi Saito and Tatsuhiko Saeki. 2013. Influential factors on electrical resistivity of hardened cementitious paste. Cement Science and Concrete Technology, Vol.66:189–196 (in Japanese).
5. M.R. Jones, D.E. Macphee, J.A. Chudek, G. Hunter, R. Lannegrand, R. Talero and S.N. Scrimgeour. 2003. Studies using ^{27}Al MAS NMR of AF_m and AF_t phases and the formation of Friedel's salt. Cement and Concrete Research, Vol.33:177–182.
6. N. Otsuki, D. Furuya, T. Saito and Y. Tadokoro. 2011. Possibility of sea water as mixing water in concrete. Proceedings of 36th Conference on Our World in Concrete & Structures, Vol.36:131–138 (in Japanese).
7. Yusuke Otsuka, Yasuhiro Dan, Nobufumi Takeda and Hidenori Hamada. 2013. Kaisui wo mochiita kouro-cement koukatai no bussei nit suite, 67th Cement gijyutsu taikai:158–159 (in Japanese).

2.2 STRENGTH

This section reports on the influence of seawater on concrete strength. In cases with only OPC and BFS, strength is higher at 7 days with seawater mixing while at the age of 20 years the influence of seawater has become small. With FA cement, strength with seawater is higher than with freshwater up to 90 days.

2.2.1 Strength of Seawater Concrete with OPC

Using various cement types, Ikebe [1] evaluated the compressive strength and flexural strength of seawater concrete. Compressive strength and flexural strength were higher up to the age of 3 days when seawater was used. However, as the material aged, this difference in compressive strength began to fall from the 7th day. By the 28th day, it had fallen below that of freshwater in almost all cases. Similarly, Akashi et al. [2] also found that until the 7th day, seawater mortar had higher flexural strength as well as compressive strength. By the 28th day, these properties were the same in both mortars, or the freshwater mortar had slightly higher values. Hasaba et al. [3] indicated that seawater concrete had higher compressive strength than freshwater concrete. When OPC was used, compressive strength was higher by 46% on the 7th day, 12% on the 28th day and then 7% on the 91st day. Seawater-mixed high early strength cement concrete has higher compressive strength, and the strength ratio does not decrease as the material ages [4].

The relationship between curing time and compressive strength index for concretes made with seawater is shown in Figure 2.4 [5]. Here, compressive strength index is the strength property of the concrete compared with that of concrete mixed with freshwater expressed as a percentage, where the concretes have the same mix design and are at the same age. Initial strength is significantly higher when seawater is used as mixing water. With regard to

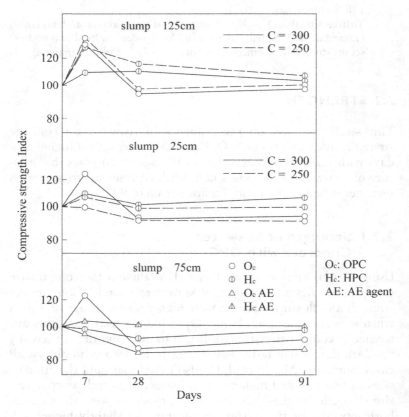

Figure 2.4 Relationship between curing time and compressive strength index [5].

Table 2.1 Compressive strength of concrete [6].

Kind of cement		Compressive strength (kg/cm²)				Seawater / freshwater			σ 10 year / σ 28 year
		28 day	1 year	5 year	10 year	28 day	1 year	20 year	
OPC₁	W	366	–	509	321	1.12	1.02	1.04	0.88
	S	409	405	518	335				0.87
OPC₂	W	–	–	524	–		1.00		–
	S	388	366	521	307				0.79
HPC	W	425	–	560	–	0.92	0.99		–
	S	390	411	555	382				0.98
MPC	W	328	–	478	344	1.09	1.00	1.10	1.05
	S	356	558	477	379				1.06
BB₁	W	380	–	563	429	1.04	0.98	0.99	1.13
	S	395	532	551	423				1.07
BB₂	W	363	–	523	389		0.98		–
	S	366	420	513	389				1.06
Alumina	W	428	–	672	526		0.90		–
	S	258	568	603	–				–
					Ave.	1.04	0.98	1.04	0.98

W: Freshwater
S: Seawater

concrete using OPC, the strength property of the seawater concrete at 7 days is 20–30% higher compared with the freshwater one. As the material ages (at 28 days and 91 days), the compressive strength index decreases. On the other hand, the strength index of high early strength cement changes only slowly even with aging; in this case, the strength property of seawater concrete at any age is better than that of freshwater concrete.

Mori et al. [6] and Fukute et al. [7] examined the compressive strength ratio over a longer period of aging, up to 20 years (Table 2.1). In this study, the types of cement tested were OPC, high early strength Portland cement (HPC), moderate heat Portland cement (MPC), blast furnace cement type B (BB), and aluminum cement. In the figure, compressive strength ratio is the strength of the concrete made with seawater/the strength of concrete made with freshwater.

This shows that when ordinary Portland cement is used, this ratio exceeds 1.1 at an early age, representing a significant strength increase. At the age of 5 years and 20 years, this ratio tends to evolve to around 1.02 to 1.03.

BIBLIOGRAPHY

1. Midori Ikebe. 1959. Comparative testing of fresh water and seawater concretes using various cements. *Cement & Concrete*, No.146:16–22 (in Japanese).
2. Toyoki Akashi, Fumio Yamaji, Yoshimi Michikiyo and Kunihiko Hidaka. 1965. Fundamental study on properties of mortar using seawater. *Cement Science and Concrete Technology*, Vol.19:233–236 (in Japanese).
3. Shigemasa Hasaba, Mitsunori Kawamura and Jiro Takakuwa. 1973. Several properties of concrete using seawater. Proceedings of Annual Conference of the Japan Society of Civil Engineers, Vol.28, No.5:207–208 (in Japanese).
4. Shigemasa Hasaba, Mitsunori Kawamura and Yutei Yamada. 1974. The properties of concrete using seawater. Proceedings of Annual Conference of the Japan Society of Civil Engineers, Vol.29, No.5:97–98 (in Japanese).
5. Shigemasa Hasaba, Mitsunori Kawamura, Yutei Yamada and Jiro Takakuwa. 1975. Several properties of concrete using seawater as mixing water. *Journal of the Society of Materials Science*, Vol.24, No.260:425–431 (in Japanese).

6. Yoshio Mori, Nobuaki Otsuki and Osamu Shimozawa. 1981. Study on the durability of concrete mixed with seawater under marine environment (ten year test). *Cement Science and Concrete Technology*, Vol.35:341–344 (in Japanese).

7. Tsutomu Fukute, Kunio Yamamoto and Hidenori Hamada. 1990. Study on the long term durability of concrete mixed with seawater as mixing water under marine environment. Proceedings of Annual Conference of the Japan Society of Civil Engineers, Vol.45, No.5:440–441 (in Japanese).

8. Tetsuro Suzuki, Osamu Kiyomiya, Toru Yamaji, Hiroshi Takenaka, Takahiro Sakai and Ryoichi Tanaka. 2012. The development of superplasticizer containing a viscosity-modifying admixture for self-compacting concrete using seawater and sea sand. Proceedings of Annual Conference of the Japan Society of Civil Engineers, Vol.67, No.5:1181–1182 (in Japanese).

9. Kazuo Yamada, Shunsuke Hanehara, Kennichi Honma and Shunkichi Sutou. 1999. Controlling of the adsorption and dispersing force of polycarboxylate-type superplasticizer by sulfate ion concentration in solution phase. *Cement Science and Concrete Technology*, No.53:128–133 (in Japanese).

10. Osamu Kiyomiya, Penta-Ocean Construction Co., Ltd., TOA Corporation, TOYO Construction Co., Ltd. and Construction Research and Technology GmbH (Osamu Kiyomiya, Kiyofumi Sano, Eiji Sueoka, Hideharu Naito, Takahiro Sakai, Atsuro Moriwake, Takashi Habuchi, Minoru Yaguchi and Yusuke Baba). Additives for cement composition with a high salt content and cement composition with a high salt content. Japanese Unexamined Patent Application Publication No.2013-142051 (in Japanese).

11. Construction Research and Technology GmbH (Yusuke Baba, Tetsuro Suzuki and Minoru Yaguchi). Additives for cement composition with a high salt content and cement composition, Japanese Unexamined Patent Application Publication No.2013-142050 (in Japanese).

12. Nobufumi Takeda, Yoshikazu Ishizeki, Shigeru Aoki and Nobuaki Otsuki. 2011. The Improvement effect of concrete strength and watertightness using seawater. Proceedings of Annual Conference of the Japan Society of Civil Engineers, Vol.66, No.5:581–582 (in Japanese).

13. Yusuke Otsuka, Yasuhiro Dan, Nobufumi Takeda and Hidenori Hamada. 2013. Physical properties of hardened blast-furnace slag cement using seawater. Proceedings of Annual Conference of the Japan Cement Association, Vol.67:158–159 (in Japanese).

2.2.2 Hardened Properties of Seawater Concrete Made with BFS Cement

Takenaka et al. [2] found that the compressive strength of self-compacting concrete using BFS cement type B (Japan Industrial Standard: JIS) and seawater (SW-SS, SW-LS) is higher at an early age compared with the basic formulation using freshwater (Figure 2.5(a)). The strength enhancement reaches about 40% by the 7th day and falls back to almost the same value by the 28th day, which means the difference reduces with the passage of time. Further, the compressive strength, static elasticity modulus, and splitting tensile strength relationships correlate uniformly regardless of the type of mixing water and aggregate, which means they are not affected by the chloride ions.

Figure 2.5(b) [8] shows an example of the strength development of BFS cement mortar using seawater when immersed in seawater. During immersion, none of the mortars at one year of age exhibited a decrease in compressive strength, but by the 28th day the BFS cement mortars had greater compressive strength than the OPC mortar, and the compressive strength of the mortar with BFS cement type C (JIS) was significantly higher at 91 days. This means that long-term compressive strength development is better if the BFS replacement ratio is larger when seawater is used as the mixing water and curing takes place immersed in seawater.

The long-term (20 years) influence of seawater on strength is shown in Table 2.1. In this table, the 7-day compressive strength ratios of BB and BB (SO_3 increased) are 1.04 and 1.01, and the 20-year ratios are 0.96 and 0.93. These ratios at 20 years of age are just below 1.00, but as they remain above 0.9 we judge there is little difference between using seawater and freshwater.

BIBLIOGRAPHY

1. Takakazu Maruyasu, Kazusuke Kobayashi and Yoshifumi Sakamoto. Study of Portland blast-furnace slag cement concrete: Report of the Institute of Industrial Science, University of Tokyo, Vol.15, No.4:31–33. (in Japanese)
2. Hiroshi Takenaka, Hideharu Naito, Takashi Habuchi and Osamu Kiyomiya. 2012. Fundamental characteristic of self-compacting concrete using seawater and sea sand. *Proceedings of the Japan Concrete Institute*, Vol.34, No.1:1912–1917 (in Japanese).

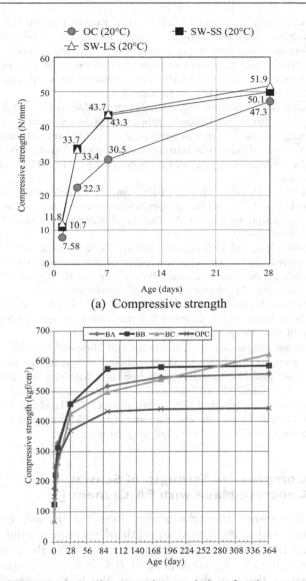

(a) Compressive strength

(b) Strength development characteristics of various cement mortars mixed with seawater and immersed in seawater

Figure 2.5 Strength development characteristics of various cement mortars [1, 2, 8].

3. Tetsuro Suzuki, Osamu Kiyomiya, Toru Yamaji, Hiroshi Takenaka, Takahiro Sakai and Ryoichi Tanaka. 2012. The development of superplasticizer containing a viscosity-modifying admixture for self-compacting concrete using seawater and sea sand. Proceedings of Annual Conference of the Japan Society of Civil Engineers, Vol.67, No.5:1181–1182 (in Japanese).

4. Yusuke Baba, Takumi Sugamata, Hayato Matsukura and Minoru Yaguchi. 2011. Fundamental properties of low-viscosity self-compacting concrete with a superplasticizer containing an innovative viscosity-modifying admixture. Proceedings of Annual Conference of the Japan Society of Civil Engineers, Vol.66, No.5:1143–1144 (in Japanese).

5. Shigemasa Hasaba, Mitsunori Kawamura, Yutei Yamada and Jiro Takakuwa. 1975. Several properties of concrete using seawater as mixing water. *Journal of the Society of Materials Science*, Vol.24, No.260:425–431 (in Japanese).

6. Hiroshi Takenaka, Takahiro Sakai, Toru Yamaji and Osamu Kiyomiya. 2013. The characteristics of self-compacting concrete using seawater and sea sand. *Proceedings of the Japan Concrete Institute*, Vol.35, No.1:1501–1506 (in Japanese).

7. Mitsunori Kawamura and Yutei Yamada. 1974. The use of seawater as mixing water. *Cement Science and Concrete Technology*, Vol.28:180–185 (in Japanese).

8. Tsutomu Fukute, Kunio Yamamoto and Hidenori Hamada. 1990. Study on the durability of concrete mixed with seawater. Report of the Port and Harbour Research Institute, Ministry of Transport, Vol.29, No.3:57–93 (in Japanese).

9. Kazusuke Kobayashi. 1978. Application of Portland blast-furnace slag cement to marine concrete structures. Technical data sheet, Nippon Steel Chemical Co., Ltd (in Japanese).

2.2.3 Compressive Strength of Seawater Concrete Made with FA Cement [I]

Figure 2.6 shows the relationship between fly ash replacement ratio and compressive strength of specimens containing NaCl. Up to a fly ash replacement ratio of 30%, the compressive strength ratio remains around 0.8–0.9, and there is no improvement through the addition of NaCl. At higher fly ash replacement ratios, the compressive strength ratio increases. In particular, when fly ash replaces 70% of the cement, the effect of NaCl addition increases with hydration age. At 28 and 91 days, a remarkable effect is seen, with compressive strength ratios reaching 1.7–1.8.

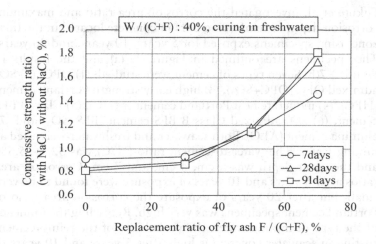

Figure 2.6 Influence of fly ash replacement ratio on compressive strength ratio.

BIBLIOGRAPHY

1. Kazuto Fukudome, Yoshitaka Ishikawa, Nobuaki Otsuki and Takahiro Nishida. 2013. Effects of NaCl on strength development of Fly-ash mixed. *JCI Proceedings of Annual Convention*, Vol.35, No.1:187–192 (in Japanese).

2.3 DURABILITY

2.3.1 Steel Corrosion in Seawater Concrete – Long-term Exposure Tests

This subsection describes some long-term exposure tests carried out on seawater concrete for periods exceeding 20 years. The results show that, over the long term, the type of cement and the wetting/drying conditions of the structure are more influential than the type of mixing water used.

(1) Long-term Exposure Test 1 – the Influence of Mixing Water versus Type of Cement in the Tidal Zone

Research on the long-term durability of seawater concrete has been conducted at the Port and Airport Research Institute (PARI).

Fukute et al. investigated the corrosion area ratio and maximum corrosion depth of reinforcing bars in concrete located in the tidal zone using specimens exposed for 5 years, 10 years, and 20 years. The specimens are outlined in Figure 2.7(a) and the results in Figure 2.7(b) Seven types of cement were studied: (1) OPC, (2) SO_3 admixed OPC (OPC+SO_3), (3) high early strength Portland cement (HPC), (4) moderate heat Portland cement (MHC), (5) Class B BFS cement, (6) SO_3 admixed Class B BFS cement (BFS+SO_3) and (7) alumina cement (ALC). Both seawater and freshwater were used as mixing water. The water-to-cement ratio (W/C) was 52.1–55.5% and the cover depth was 2 cm, 4 cm, and 7 cm. Corrosion area ratios after 5 years and 10 years of exposure were found to be very small, but after 20 years of exposure the corrosion area ratio of Portland cement specimens was very high. Regarding the influence of the type of mixing water, the probability of the reinforcement rusting in seawater concrete is high after 5 years and 10 years of exposure, but the results for all cases were similar after 20 years of exposure. Further, with regard to maximum corrosion depth, there was no confirmed influence of the type of mixing water. These results show that the type of mixing water has an influence during

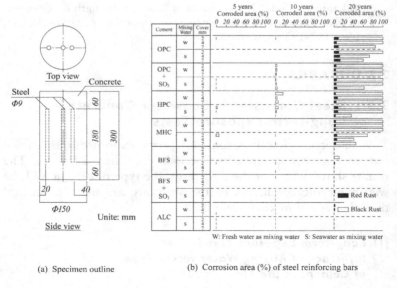

(a) Specimen outline   (b) Corrosion area (%) of steel reinforcing bars

Figure 2.7 Specimen outline and corrosion area (%) of steel reinforcing bars.

early exposure when corrosion is slightly high, but there is little influence after long-term exposure. In contrast, cement type has a considerable influence on corrosion.

(2) Long-term Exposure Test 2 – the Influence of Mixing Water versus Wet and Dry Condition in the Tidal Zone

Akira et al. [4] and Yonamine et al. [5] investigated steel corrosion in concrete after 26 years of exposure in a marine tidal zone. They used five kinds of cement: (1) OPC, (2) Class A BFS cement, (3) Class B BFS cement, (4) Class C BFS cement, and (5) Class B FA cement. Seawater and freshwater were used as the concrete mixing water. W/C was 45% and 55% and the cover depth was 2 cm, 4 cm, and 7 cm. Figure 2.8 shows the results obtained at three long-term exposure tests (tidal pools) at PARI and Kagoshima port. Each exposure site had different conditions. In this figure, differences of cement type and cover depth are included. Defining the immersion ratio as the ratio of wet exposure time (time spent in seawater) to total exposure time, the ratio was 0.34 in 'Tidal pool–15 years,'

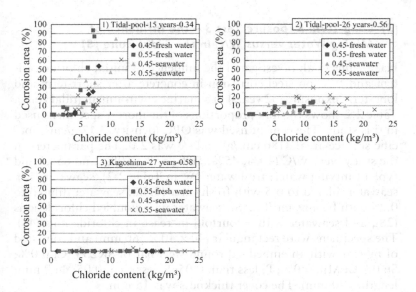

Figure 2.8 Relationship between chloride ion concentration at reinforcing bar surface and corrosion area ratio.

0.56 in 'Tidal pool–26 years' and 0.58 in 'Kagoshima port–27 years.' The specimens exposed at Kagoshima port suffered heavy fouling by marine organisms and were fitted with repellent coverings; as a result, these specimens remained in a highly wet condition not conducive to drying, so close to the submerged condition. Regardless of cement type, the corrosion area ratio tended to increase with increasing chloride ion concentration at the steel surface. The corrosion area ratio for 'Tidal pool–15 years' (immersion ratio: 0.34) was large, but no corrosion was observed in specimens exposed at 'Kagoshima–27 years' (immersion ratio: 0.56), which is assumed to be an environment similar to submersion in seawater. That is, chloride ion concentration did influence corrosion in the 'Tidal pool–15 years' and 'Tidal pool–26 years' cases; however, in the case of 'Kagoshima–27 years' it had no influence.

It is inferred that the dissolved oxygen concentration around the reinforcing bars had a significant influence on corrosion. Further, the limit chloride ion concentration for corrosion was about 2 kg/m^3 in the case of 0.34 immersion ratio, about 4 kg/m^3 in the case of 0.56 immersion ratio, and 10 kg/m^3 or more in the case of 0.58 immersion ratio.

(3) Long-term Exposure Test 3 – the Influence of Mixing Water versus W/C in the Tidal Zone [8]

Hashizume et al. investigated the influence of initial chloride content and W/C on steel corrosion in concrete after short-term and long-term exposure (32 years) in a marine environment (tidal zone). Table 2.2 shows the mix proportions of the mortar specimens used in this study. The cement used was OPC (density: 3.17 g/cm^3; specific surface area: 3180 cm^2/g) and s/c was 2.0. The parameters in the study were W/C (40%, 45%, 50%, 55%, 60%, and 65%) and type of mixing water: freshwater (W in Table 2.2), seawater (1S), seawater diluted to 0.5 with freshwater (0.5S), seawater diluted to 0.25 with freshwater (0.25S), seawater with doubled chloride ions (2S), and seawater with a fourfold increase in chloride ions (4S). The specimens were rectangular (40 × 40 × 160 mm) and consisted of mortar with an embedded round steel bar of SR235 (C: 0.0%; Si: 0.0%; Mn: 0.0%; P: less than 0.05; S: less than 0.05; Φ: 9 mm; length: 100 mm). The cover thickness was 15 mm.

In this study, corrosion weight loss was evaluated by measuring the difference between the weight of the steel bar in the specimen

Table 2.2 Mortar mix proportions

Specimens Mark	W/C (%)	Cement (kg/m³)	Water (kg/m³)	Sand (kg/m³)	Initial chloride content (%)
40-W	40	674.9	270.0	1439.8	0
40-1S					0.2
40-2S					0.41
40-4S					0.82
45-W	45	652.9	293.8	1305.7	0
45-1/4S					0.06
45-1/2S					0.11
45-1S					0.23
45-2S					0.45
50-W	50	632.9	316.1	1264.5	0
50-1S					0.07
55-W	55	612.9	337.1	1225.7	0
55-1/4S					0.07
55-1/2S					0.14
55-1S					0.27
55-2S					0.54
55-4S					1.08
60-W	60	594.6	356.7	1189.3	0
60-1S					0.29
65-1/4S	65	582.2	378.4	1164.3	0.08
65-1/2S					0.16
65-1S					0.31
65-2S					0.62

and the calculated initial weight of the bar. Figure 2.9 shows the influence of W/C and initial chloride content on corrosion loss after 32 years of exposure in the marine tidal zone. This confirms the influence of W/C on corrosion loss. With higher W/C, the corrosion weight loss is greater. On the other hand, there was no evidence of initial chloride content influencing corrosion loss.

These results are suitable for multiple regression analysis, a method of multivariate statistics that shows the influence on an objective variable using explanatory variables in the form of a 'T' value. T values for the influence of initial Cl⁻ content and W/C are −0.30 and 2.73, respectively. Generally, a T value exceeding 2.0 represents a major effect, so these results show that W/C has a

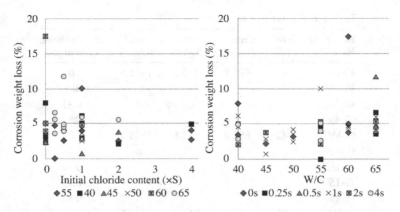

Figure 2.9 Corrosion weight loss due to W/C and initial chloride content after 32 years of exposure.

major effect on corrosion weight after 32 years of exposure in the tidal zone, while initial Cl⁻ content is not a major factor.

(4) Summary of Steel Corrosion in Seawater Concrete

The type of mixing water has a small influence on the corrosion of steel in concrete exposed to the tidal zone during initial exposure up to about ten years. After long-term exposure, this influence becomes very small, while the type of cement used has a great influence on corrosion. However, in seawater concrete, the chloride ion concentration exceeds the corrosion limit because the initial chloride ion concentration is estimated to be about 3.0 kg/m³. This means that corrosion of the steel is likely to progress gradually from the initial time. Corrosion of steel in reinforced concrete structures mixed and exposed to the splash zone and the atmospheric zone is expected to be higher when seawater is used compared with freshwater, assuming all other materials are the same.

These results demonstrate that, in terms of steel corrosion, seawater concrete does not pose a problem for undersea applications. In other situations, particularly in areas poor in freshwater resources where the use of seawater cannot be avoided, the sea can be considered an effective source of mixing water as long as the materials and mix proportions are chosen for high resistance to chloride attack. Seawater concrete is expected to satisfy these performance requirements when a cement highly resistant to chloride attack is used, such as Class B BFS cement, along with a reduced water-to-cement ratio. If steel corrosion cannot be permitted at all, even over long-term service, the use of a highly corrosion-resistant reinforcement, such as stainless steel bars or epoxy-coated bars, would be desirable.

BIBLIOGRAPHY

1. H.H. Uhlig and R.W. Revy. 1999. Corrosion reaction and control, Industrial Library (in Japanese).
2. Japan Society of Civil Engineers Concrete Committee.2012. Standard specifications for concrete structures. Japan Society of Civil Engineers (JSCE), (in Japanese).
3. Fukute, T., Yamamoto, K. and Hamada, H. 1990. Study on durability of concrete mixed with seawater and exposed to marine environment. Report of Port and Harbor Research Institute, Vol.29, No.3:57-93 (in Japanese).
4. Akira, Y., Yamaji, T., Kobayashi, H. and Hamada, H.. 2012. Long-term durability of concrete mixed with seawater in tidal zone. *Proceedings of the Japan Concrete Institute*, Vol.34, No.1:820–825 (in Japanese).
5. Yonamine, K., Yamaji, T., Kobayashi, H., Akira, Y. 2013. Corrosion state of steel bars in concrete in different tidal zones. Proceedings of the Concrete Structures Scenarios, Vol.13:77-82 (in Japanese).
6. JSCE. 2012. Research Committee report of steel corrosion and anti corrosion, (338) Part 2, No.99 (in Japanese).
7. Maruya, T., Takeda, K., Horiguchi, K., Oyama, T. and Hsu, K.L. 2006. Simulation of steel corrosion in concrete based on macro-cell corrosion circuit model. *Journal of JSCE*, Vol.62, No.4 (in Japanese).
8. Koki Hashizume, Ayako Mizuma and Nobuaki Otsuki. 2015. Influence of initial chloride content and water-cement ratio on corrosion of steel bar in mortar exposed to marine environment for 32 years. Seminar-Workshop on the Utilization of Waste Materials, 'Science and Engineering on Waste Utilization for the People, Economy and the Environment,' Henry Sy Sr. Hall, De La Salle University.

2.3.2 Freeze–Thaw Resistance

This subsection considers the influence of freeze–thaw action on OPC and BFS concretes made with seawater, especially in respect of the concrete's dynamic modulus. The results presented demonstrate that the dynamic modulus of seawater concrete after 300 freeze–thaw cycles is acceptable if there is sufficient entrained air content. This means that freeze–thaw action presents no particular problems for seawater concrete.

In concrete containing chloride ions that is exposed to freeze–thaw action, a higher chloride ion content leads to an increase of scaling and decrease in relative dynamic elasticity [1–4]. Hashimoto et al. proposed as an explanation for this that, in the case of concrete exposed to chloride ion ingress owing to spraying with an anti-freezing agent, monosulfate decomposes into ettringite resulting in loss of tensile strength and acceleration of freeze–thaw damage [5]. They also reported that, by replacing OPC with BFS, it is possible to suppress the formation of the ettringite that contributes to frost damage [6]. However, reports on the amount of chloride ions required in the concrete to adversely affect its freeze–thaw resistance are still few in number. According to Yamato et al., with concrete (W/C: 55%) mixed using that contains chloride ions in the proportion of 0.5% NaCl by dry weight of sand (NaCl: about 2.5 kg/m^3), if the amount of air entrained by an AE agent is assumed to be on the order of 6% (approximate bubble spacing factor of 200 μm), an 80% relative dynamic elastic modulus after 300 freeze–thaw cycles can be achieved as shown in Figure 2.10(a). However, beyond 0.5% chloride ions by dry weight of sand, a sudden drop in relative dynamic elastic modulus is reported [7].

Takeda et al. carried out freeze–thaw tests with OPC concrete (W/C: 50%) containing chloride ions in the mixing water as shown in Figure 2.10(b). For chloride ion contents up to 2.5 kg/m^3, a relative dynamic elastic modulus of more than 90% can be ensured at 200 cycles with an air content 4.5% or more, but when the chloride ion content is 5 kg/m^3, the modulus of elasticity drops sharply [8]. Figure 2.10(c) shows the relative dynamic elastic modulus of seawater concrete (BFS 50% substitution; W/C: 50%; chloride ion content: 3.1 kg/m^3) in freeze–thaw tests. These results show that if the air content of the concrete is 5.5% or higher, the relative dynamic elastic modulus after completion of 300 cycles is approximately 95% with the use of seawater as the mixing water, even

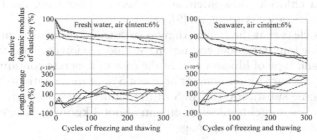

(a) Durability of concrete including chloride ions under freeze-thaw action
(Cement: OPC; W/C: 55%; air content: 6%)

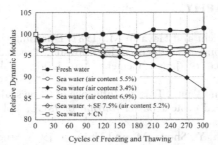

(b) URelative dynamic modulus of concrete containing chloride ion under freeze-thaw
conditions (Cement: OPC; W/C: 50)

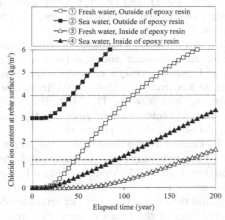

(c) Relative dynamic modulus of concrete using seawater under freeze-thaw conditions (Cement: BB; W/C: 50%; chloride ion content: 3.1 kg/m3) [9]

Figure 2.10 Durability and Relative dynamic modulus of concrete under
freeze–thaw action [7–9].

with a chloride ion content of 3.1 kg/m^3. Therefore, the freeze–thaw resistance is adequate [9].

It is clear that the freeze–thaw resistance of concrete containing chloride ions varies depending on the chloride ion content, air content, and type of binder used. In the case of concrete containing an AE agent with an air content of about 5% or more with 50% W/C and some substitution of BFS in the binder, even concrete mixed with seawater (chloride ion content: about 3 kg/m^3) can be considered to suppress frost damage.

BIBLIOGRAPHY

1. Ayuta Koichi and Hayashi M.. 1981. Durability under frost damage of concrete influenced by seawater. Proceedings of Annual Conference of Japan Cement Association, Vol.35 :325-328(in Japanese).
2. Fujii, T. and Fujita, T. 1895. Influence of Cl on scaling deterioration of hardened cement paste. Journal of JSCE, Vol.360:129-138 (in Japanese).
3. Tsukinaga, Y., Shoya, M. and Hara, T. 1993. Study on frost damage of concrete influenced by Cl. Journal of Cement and Concrete Research, No.47:129-138 (in Japanese).
4. Miura, H., Itabashi, H. and Iwaki, I. 1991. Experimental Study on Frost damage of concrete influenced by anti-freezing agent:159-166 (in Japanese).
5. Hashimoto, K., Yokota, H., Sato, Y. and Sugiyama, T. 2011. Influence of frost damage and anti-freezing agent on hydration products and tensile strength characteristics. Journal of Cement and Concrete Research, Vol.65:400-405 (in Japanese).
6. Hashimoto, K. and Yokota, H. 2012. Pore structures and Cl-immobilization of cement matrix. Journal of Cement and Concrete Research, Vol.66:460-465(in Japanese)
7. Yamato, T., Emoto, Y. and Soeda, M. 1984. Freezing and thawing resistance of concrete made with seashore sand. Proceedings of the Japan Concrete Institute, Vol.6:233–236. (in Japanese)
8. Takeda, N. and Sogo, S. 2001. Influence of combined deterioration of freezing and thawing action and salt attack on durability of concrete. Proceedings of the Japan Concrete Institute, No.2:427–432 (in Japanese).
9. Keisaburo Katano. 2013. Properties and application of concrete made with seawater and un-washed sea sand. SCMT3. http://claisse.info/2013%20papers/data/e172.pdf

2.3.3 Alkali–Silica Reaction

This subsection considers the effect of concrete mixing seawater and absorbed seawater on the alkali–silica reaction (ASR) in concrete. The results show that mixing seawater promotes ASR and that an appropriate countermeasure is to use BFS or FA as a mineral admixture. The mechanism by which the mixing seawater affects ASR is different from that when seawater is absorbed into the concrete from an external source [1]. These two different mechanisms are discussed below.

(1) Case 1: Seawater as Mixing Water in Concrete

It is natural to ask whether the use of seawater as the mixing water in concrete will promote ASR. The presence of sodium and potassium ions from seawater at a constant concentration can be expected to result in an increase in the pH of the pore solution, thereby promoting ASR [1–3].

Kawamura [1] stated that the NaCl present in seawater used as mixing water reacts with C_3A ($3CaO \cdot Al_2O_3$) during cement hydration, increasing the OH ion concentration of the pore solution and facilitating the alkali–silica reaction:

$$C_3A + 2NaCl + Ca(OH)_2 + 10H_2O \rightarrow C_3A \cdot CaCl_2 \cdot 10H_2O$$

$$+ 2Na^+ + 2OH^-$$

An experiment [2] in which NaOH and NaCl were separately added in equal amounts showed similar expansion due to the alkali–silica reaction. NaOH and NaCl are thought to produce equivalent sodium ion contents leading to similar pH conditions in the pore solution. Hence NaCl is considered to have a profound effect on the promotion of ASR. Since Na and K ions are present at a constant ratio in seawater, the amount of alkali required to suppress ASR, which is generally indicated as 3 kg/m³, may exceed the regulated value. Therefore, when using potentially reactive aggregate, even if only slight expansion occurs at an early age when freshwater is used as mixing water it must be assumed that expansion will be greater in the case of seawater.

In using seawater in a concrete mix, the nature of the reactive aggregate, the types of binder, the exposure environment, and many other factors should be considered in addition to practicality and standardization. To this end, research data needs to be accumulated on the practical use of seawater in concrete. In the meantime, if the concrete is produced using seawater as mixing water, appropriate ASR control measures should be taken, such as using harmless aggregates or ensuring the use of the specified cement. If the seawater used in concrete contains considerable alkali content (approximately 5 kg/m^3 alkali (Na_2O equivalent)), the use of BFS or FA can be expected to suppress ASR [1, 3, 4]. Note that the literature [3] provides a calculation method for the amount of admixture (BFS) needed to inhibit expansion when using seawater as mixing water.

(2) Case 2: Seawater from External Source

Many reports have confirmed that ASR is promoted by seawater penetrating from outside. The mechanism for this is given by Kawamura [1] and the mechanism is given as follows. Na and K enter the concrete from the seawater, causing an acceleration of ASR as in Case 1. Meanwhile, since the concrete is in contact with seawater, the OH⁻ ions in the concrete (water) diffuse out from inside, so the OH⁻ concentration decreases. Since both phenomena proceed simultaneously, if seawater is supplied from the outside, ASR is not necessarily promoted. Other investigations of structures in marine environments have been conducted [4–6], showing evidence of ASR promotion when seawater is supplied from the outside.

It is clear that the influence of seawater on the alkali-aggregate reaction differs according to whether it is supplied externally or included as mixing water. If seawater is used as mixing water and the aggregate is potentially reactive, concrete expansion needs to be verified in advance and it is necessary to use BFS or FA. With the use of seawater, concrete being contemplated in Southeast Asia, Africa, and other regions in the future, it is necessary to ensure sufficient knowledge of ASR in these regions beforehand. Recently, ASR has become a problem and a major concern in Thailand [7]. In the future, experience and learning from dealing with this problem will become applicable to ASR in seawater concrete.

BIBLIOGRAPHY

1. Kawamura, M. and Chatterji, S. 2002. *Material Science of Concrete.* Morikawa Press (in Japanese).

2. Kawabata, Y., Yamada, K. and Matsushita, H. 2007. Petrological study on evaluation of alkali-silica reactivity and expansion analysis of andesite. *Journal of JSCE*, Vol.63, No.4 (in Japanese).
3. JCI. 2014. Technical Committee Report on Diagnosis of ASR-affected Structures (in Japanese).
4. Torii, K. et al. 2002. Investigation on damage caused by ASR in RC bridges where de-icing salts are used. *Proceedings of the Japan Concrete Institute*, Vol.24, No.1:579–584 (in Japanese).
5. JCI. 1999. Technical Committee Report on Degradation of concrete due to Anti-Freezing Agent (in Japanese).
6. Habuchi, T. and Torii, K. 2004. Deterioration of concrete and evaluation method of combined effect of ASR and seawater, *Journal of JSCE*, Vol.774:V–65 (in Japanese).
7. Yamada, K., Hirono, S. and Ando, Y. 2013. ASR problems in Japan and advice on ASR problems in Thailand, *Journal of Thailand Concrete Association*, Vol.1, No.2:1–19.

(3) ASR Prevention by Mineral Admixtures

As described in former sections, the question 'Does seawater accelerate ASR in concrete?' is still in seawater concrete usage. The answer to this question is summarized in the two studies cited below. When seawater concrete is made using OPC only, then ASR readily causes expansion in the presence of reactive aggregate. However, if an adequate volume of mineral admixtures such as BFS or FA are included, ASR expansion does not occur even with a reactive aggregate. So to prevent ASR in seawater-mixed or cured concrete, the key is using a mineral admixture [1, 2].

BIBLIOGRAPHY

1. Adiwijaya. 2015. *A fundamental study on seawater concrete related to strength, carbonation and alkali-silica reaction*, PhD Thesis, Kyushu University.
2. Yuichiro Kawabata, Takashi Habuchi, Jun Kutsuna and Kazuhide Yonamine. 2017. Inhibiting effect of Blast Furnace slag on ASR expansion of seawater concrete. *Cement Science and Concrete Technology*, Vol.71: 323-330.

2.3.4 Long-term Performance Change – Long-term Exposure Test [1]

A paper published in 1971 [1] was a great inspiration to this book's editors, and particularly to Nobuaki Otsuki, who began his research on concrete in 1972 as a senior student at the Tokyo Institute of Technology. This paper by Odd E. Gjorv is summarized briefly here.

(1) Outline

This long-term exposure test was planned between 1938 and 1943 by Prof. Olav Heggestad and his assistant H. M. Bakken at the Technical University of Norway. More than 2,500 concrete test specimens were exposed to seawater at temperatures of 1°C to 12°C at a station near Trondheim Harbor. In 1962, the Norwegian Committee on Concrete in Seawater was asked to complete the tests and present to the profession any significant results that could still be salvaged. Once the entire test had been completed, Prof. Odd E. Gjorv wrote the paper to briefly present some of the results. Here, the results relating to seawater concrete are summarized.

(2) Material Used

Four series of specimens were reported altogether. Of these, Series 4 was a study of seawater versus freshwater as mixing water. The mixing proportions by weight were 1.0 (water):3.08 (cement):3.29 (sand), and the cement content was 312 kg/m³. The water-to-cement ratio was approximately 0.60 and the consistency measured in terms of the slump was 8 cm.

(3) Fabrication and Exposure of Specimens

All specimens were made of plain concrete and were cast in the laboratory. Forms were removed after two days of curing under wet conditions. Then most of the specimens were directly exposed to seawater. The majority of specimens were placed in a rock basin measuring 8 m in length, 3 m in width, and 3 m in depth. A pump circulated 150 m³ of fresh seawater through the basin every day from a depth of 30 m. All specimens at the station were continuously immersed. The salinity of the water was fairly constant at 3.0%.

(4) Test Results and Conclusions (Series 4)

Eighteen specimens of each series were prepared. Tests were performed seven times in the first 15 years, with two specimens tested at a time. The remaining four specimens (18 − (2 × 7)) of each were tested after 30 years (though some were tested after 25 years). Three tests of flexural strength and four tests of compressive strength were performed on each concrete prism (10 × 10 × 75 cm), using successively broken parts.

In Series 4, the effect of seawater as mixing water was examined for concretes made with both high alumina cement and Portland cement. With a cement content of 313 kg/m^3, no significant difference in durability was found with the two tested cements. That is, the use of seawater versus freshwater as mixing water was found to have no significant effect on the behavior of the concrete, whether Portland cement or high alumina cement concrete.

BIBLIOGRAPHY

1. Odd E. Gjorv. 1971. Long-term durability of concrete in seawater. *ACI Journal*, Vol.68:60–67.

Chapter 3

Special Techniques for Seawater Concrete

3.1 INTRODUCTION

This chapter outlines the important considerations to be made in the construction of durable structures using seawater concrete and a number of special techniques that can be applied. Regarding concrete mix design, W/C (or the water-binder ratio, W/B) should be less than 50% and certain mineral admixtures should be added. The use of a corrosion inhibitor is recommended to prevent corrosion of embedded rebars as well as the use of reinforcement that is not susceptible to corrosion, such as stainless steel bars, epoxy-coated steel bars, bamboo, or FRP. Performance-based design is a suitable tool for designing seawater concrete. Also, the use of chemical admixtures is an effective means of extending the life of seawater concrete.

3.2 MIX DESIGN

3.2.1 W/C of Seawater Concrete

Use of a W/C less than 50% is strongly recommended. W/C is one of the main factors determining the durability of concrete structures and sufficient durability can be obtained with W/C of less than 50%.

3.2.2 Materials Selection

The use of an aggregate that is non-reactive in ASR is strongly recommended. A sufficient amount of mineral admixtures such as

DOI: 10.1201/9781003194163-3

ground granulated blast furnace slag (GGBFS), fly ash (FA), shirasu, or similar should be added.

3.3 PREVENTION OF STEEL CORROSION IN SEAWATER CONCRETE

This section discusses methods of reducing steel corrosion in seawater concrete. Some of the available corrosion inhibitors and types of non-corroding reinforcement (including strands and wires) are covered. Certain inhibitors and reinforcement materials are recommended.

3.3.1 Concrete Cover

Along with W/C, the provision of sufficient concrete cover is crucial to ensuring the durability of seawater concrete against steel corrosion. Seawater concrete contains a certain amount of chloride ions, making it more vulnerable to corrosion than concrete mixed with freshwater. However, once steel bar corrosion has begun in concrete, it is the supply of oxygen that is the main determinant of corrosion rate. A greater cover depth prevents oxygen from penetrating as deep as the steel reinforcement, so the corrosion rate will be lower.

3.3.2 Corrosion Inhibitor

One way to prevent corrosion of steel reinforcement in seawater concrete is to use a corrosion inhibitor. Sabrina Harahap, a former master's student at Kyushu University, started an experimental study of this method [1]. The experimental work to evaluate effectiveness is still being carried out by the following students. However, at least until two years into the test, pre-coating the steel reinforcement with a mortar, including an inhibitor is effective against premature steel corrosion.

BIBLIOGRAPHY

1. Sabrina Harahap. 2019. *A study on corrosion behaviour of steel reinforcement in seawater-mixed concrete and application of corrosion prevention: Corrosion inhibitor and cathodic prevention.* Master's Thesis of Graduate School of Kyushu University.

3.3.3 Types of Reinforcement

In this subsection, the use of corrosion-resistant reinforcement in seawater concrete is discussed. Although few cases of actual application of less/non-corroding reinforcement in seawater concrete have been reported, it is considered that for reinforcement to work effectively in seawater concrete, appropriate materials should be chosen.

(1) Stainless Steel Reinforcing Bars

Stainless steel is a metal alloy that contains 10.5% or more of Cr in Fe. It has greater resistance to corrosion than ordinary steel due to a thin layer (passive film) of Cr oxide that forms in the air. In Japan, a JIS standard (JIS G 4322) and design guidelines for stainless steel reinforcing bars have been established since 2008 to promote the practical use of stainless steel in concrete [1, 2]. The types of stainless steel covered by the standard are shown in Table 3.1. There are three types: SUS304-SD austenitic stainless steel (type 304) with 18 mass% of Cr and 8 mass% of Ni; SUS316-SD stainless steel (type 316) including Mo to give higher resistance to corrosion than the other stainless steels; and SUS410-SD ferritic or martensitic stainless steels (type 410) with 12 mass% of Cr as low-cost stainless steel [1].

The corrosion threshold value of chloride ions is higher for stainless steel than for ordinary steel (1.2 kg/m^3 according to the JSCE standard): SUS410-SD is 7.5 times higher, SUS304-SD is 13 times higher, and SUS316-SD is 70 times higher. Therefore, it can be said that the use of stainless steel can increase the durability of reinforced concrete structures, achieve a long design life, and reduce the lifetime cost of structures [2]. Stainless steel also has high resistance against corrosion due to the carbonation of concrete [6]. Moreover, it is reported that corrosion due to macro-cell occurred between ordinary steel and stainless steel is smaller than that between un-corroded ordinary steel and corroded steel, although inter-metal corrosion is a concern [3].

Type 304 and type 316 stainless steels are usually used in countries other than Japan. The oldest example is a concrete bridge (constructed 1937 to 1941) built using type 304 stainless steel in Progresso, Mexico. Recently Stonecutters Bridge was constructed in Hong Kong with type 304 stainless steel. The need for steel that offers high resistance against corrosion is demonstrated by the

Table 3.1 Stainless steel bars (JIS G 4322) and recommended threshold value of chloride ion concentration

No	Type designation	Corresponding stainless steel grades		Strength classes	0.2% proof stress (MPa)	Tensile strength (MPa)	Recommended value of threshold chloride ion concentration[2] (kg/m³)
1	SUS304-SD	SUS304	(Austenitic)	295A	295 or more	440~660	15
		SUS304N2		295B	295~390	440 or more	
				345	345~440	490 or more	
				390	390~510	560 or more	
2	SUS316-SD	SUS316	(Austenitic)	295A	295 or more	440~660	24
		SUS316N		295B	295~390	440 or more	
				345	345~440	490 or more	
				390	390~510	560 or more	
3	SUS410-SD	SUS410L	(Ferritic)	295A	295 or more	440~660	9
		SUS410	(Martensitic)	295B	295~390	440 or more	
				345	345~440	490 or more	
				390	390~510	560 or more	

high demand for epoxy-resin-coated steel in the domestic Japanese market. We expect the use of stainless steel reinforcing bars to further increase in the future (Table 3.1).

BIBLIOGRAPHY

1. Japan Industrial Standard. 2008. JIS G 4322 Stainless Steel bars for concrete reinforcement (in Japanese).
2. JSCE. 2008. Concrete library 130 – Guideline of design and construction of reinforced concrete, Maruzen (in Japanese).
3. Yu Tadokoro and Nobuaki Otsuki. 2013. Corrosion caused by contact between stainless and normal steel in concrete. *Zairyo to Kankyou*, No.60:291–294 (in Japanese).
4. Hirofumi Umemoto. 2012. Construction of Nouoohashi bridge – first application in Japan of stainless steel in bridge girder. *JSSC*:23–26 (in Japanese).
5. Japan Concrete Institute. 2013. Practical guideline for investigation, repair and strengthening of cracked concrete structures:162.
6. Yu Tadokoro, Masao Kojima and Nobuaki Otsuki. 2010. Corrosion behavior of stainless steel in concrete. *Zairyo to Kankyo*, Vol.59:179–186 (in Japanese).
7. Yu Tadokoro, Yui Tsukuda, Toru Yamaji, Tsuyoshi Maruya and Jyunichiro Niwa. 2009. Experimental study on corrosion threshold value of chloride ions for stainless steel in concrete. *Journal of Materials, Concrete Structures and Pavements E*, Vol.65, No.4:522–529 (in Japanese).
8. JSCE. 2008. Concrete Library 130 – Guideline of design and construction of concrete reinforced concrete, Maruzen (in Japanese).
9. Ryoichi Tanaka, Toru Yamaji, Yoshikazu Akira, Osamu Kiyomiya, Takahiro Sakai and Minoru Yaguchi. 2013. A study on corrosion property of steel bars in self-compacting concrete using seawater and sea sand. NACE International East Asia & Pacific Rim Area Conference & Expo 2013, EAP13-4565.

(2) Epoxy-coated Steel Reinforcing Bars

Hoshino et al. have reported on the results of tests in which epoxy-coated steel reinforcement was exposed in the marine splash zone of an actual marine environment [1]. After five years of exposure, a steel bar coated with 200 µm of epoxy resin was corroded, but no corrosion occurred in the case of coating with 300 µm of epoxy resin. Hoshino et al. also exposed concrete specimens containing epoxy-coated steel in a marine environment for 15 years. Observation of the corrosion status after this time showed that the corroded area of the epoxy-coated steel bar was 1–2%, although the chloride ion concentration around the steel bar reached as high as 10 kg/m³. Therefore, they concluded that epoxy-coated steel offers good resistance against corrosion in concrete that has a high concentration of chloride ions [2].

Predictions of chloride ion concentration in seawater concrete and normal concrete at the resin surface and at the steel surface under the epoxy coating are shown in Figure 3.1. These predictions are based on the design guidelines published by JSCE. Exposure conditions were the splash zone (surface chloride concentration of 13 kg/m³), and BFS cement was used at a W/C ratio of 50%. The cover thickness was 70 mm. Based on these results, it is considered that the chloride ion concentration at the steel surface will exceed 1.2 kg/m³ after 170 years in the case of concrete mixed

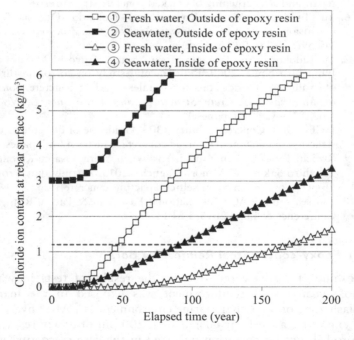

Figure 3.1 Prediction of chloride ion penetration into concrete with epoxy-coated rebar in marine environment.

with freshwater (Δ) and 90 years in the case of concrete mixed with seawater (▲). This review of the literature shows that epoxy-coated steel can be effective as reinforcement in seawater concrete structures.

In using epoxy-coated steel in concrete mixed with seawater, the possibility of corrosion developing around defects such as pin holes in the epoxy coating resulting from the manufacturing process or damage during construction is a concern. Hoshino et al. conducted five-year exposure tests of concrete specimens containing epoxy-coated steel reinforcement with defects (pinholes and defects measuring 5 mm × 5 mm). Corrosion around the defective areas was not confirmed even though chloride ion concentration in the concrete around the steel bar was 7 kg/m^3 [1]. Miura et al. studied corrosion at defect locations caused by tension or bending of the epoxy coating around the steel. In this case, the concrete contained 11 kg/m^3 of initial chloride and the specimen was exposed to dry and wet conditions to cause accelerated corrosion. The test period was equivalent to 20–30 years of exposure in an actual environment. In this case, slight corrosion was detected at the defect locations but no cracks were observed in the concrete [3]. Amry D. and Hamada et al. also investigated the corrosion of epoxy-coated steel reinforcement in concrete mixed with seawater. They reported that corrosion around defects was quite slight, even though the chloride concentration around the steel bar was 12 kg/m^3 [4].

Based on the above results, it is considered that epoxy-coated steel offers durability against reinforcing bar corrosion even if the epoxy coating has some small defects and if the initial chloride content of the concrete is as high as 3–5 kg/m^3.

BIBLIOGRAPHY

1. Tomio Hoshino and Taketo Uomoto. 2004. Research on anti-corrosion effect of epoxy coated steel. *Proceedings of Japan Concrete Institute*, Vol.26, No.1:891–896 (in Japanese).
2. Tomio Hoshino and Taketo Uomoto. 1998. Durability of RC beams with epoxy-coated bars and galvanized bars exposed in marine environment for 15 years. *Journal of Materials, Concrete Structures and Pavements*, No.592/V-39:107–120 (in Japanese).
3. Nao Miura, Hirofusa Itahashi and Tetsuzo Arai. 1988. Research on rust generated near defects in epoxy-coated steel. *Proceedings of Japan Concrete Institute*, Vol.10, No.2:523–528 (in Japanese).

4. Amry Dasar, Hidenori Hamada, Yasutaka Sagawa and Rita Irmawaty. 2013. Corrosion evaluation of reinforcing bar in seawater mixed mortar by electrochemical method. *Proceedings of Japan Concrete Institute*, Vol.35, No.1:889–894.

(3) Bamboo Reinforcement

Bamboo is a plant, so it does not suffer from the so-called corrosion phenomenon related to oxidation. If bamboo could be used as a reinforcing material, the effects of salt in seawater concrete would not need to be considered. The main issues to be considered are the possible decomposition of bamboo caused by moisture and whether detrimental bacteria can breed or not under the highly alkaline conditions found within concrete. There are three major requirements for bacteria to breed: nutrients, water, and appropriate temperature. Concrete is thought to have insufficient water for bacteria to breed under ordinary conditions, so bamboo reinforcement would be protected from bacterial attack. However, a variety of views are expressed in the literature and there is no clear answer at present as to whether bamboo reinforcement would suffer from bacterial attack.

Looking at the physical properties of bamboo, such as strength, quality control, and fracture mode, it would be difficult for bamboo to replace steel as reinforcement in concrete. More realistic would be to consider the characteristics of bamboo and find appropriate uses for bamboo-reinforced concrete. It is a fact that bamboo was used as a concrete reinforcement during World War II because of a lack of steel. However, papers and reports on the use of bamboo as reinforcement in the years since Japan's period of post-war high economic growth began are quite few in number. On the other hand, a number of reports about the use of bamboo as reinforcement have come out recently owing to the growing recognition of the importance of sustainable development.

Recent papers and reports related to bamboo as reinforcement can be categorized as follows.

A) Investigations of past use of bamboo-reinforced concrete in the remains of ancient structures [1].
B) Examination of the possibility of using bamboo as a substitute for reinforcing rods in structures [2–7].
C) Combined use of bamboo with new materials [8–13].
D) Other applications [14–16].

Though it falls outside the scope of this book, there have also been studies on the use of bamboo as short fibers [17].

While many varieties of bamboo exist, few are suitable for use as a structural material in concrete. These are limited to varieties such as Moso-dake bamboo and giant timber bamboo. It is said that the tensile strength of giant timber bamboo with nodes is 105 N/mm² and that of giant timber bamboo without nodes is 119 N/mm² [3]. Under tensile testing, bamboo fractures in the elastic state. The tensile strength of bamboo is 38% that of steel reinforcement, but its elastic modulus is very low, just 6% that of steel. The reported strength characteristics of bamboo vary with the researcher, so these data should be analyzed statistically. Examples of strength tests on bamboo are shown in Figure 3.2.

The durability of bamboo-reinforced concrete can be investigated by tracing any extant old concrete structures built during World War II. To this end, Egusa et al. conducted a field survey of bamboo-reinforced concrete using electromagnetic radar. Their survey did not find any decomposition of bamboo [1]. According to another investigation related to the long-term durability of bamboo-reinforced concrete, some bamboo was found to have decomposed due to poor construction.

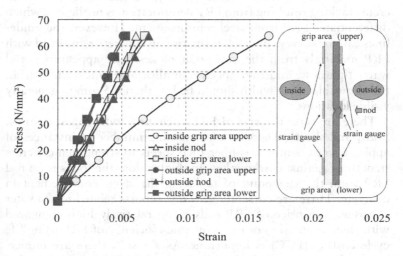

Figure 3.2 **Stress-strain curve of Moso-dake bamboo (comparison between inside and outside) [3].**

Instances of application to reinforced concrete in marine environments are very few, although the number of papers related to bamboo reinforcement has recently increased in the fields of civil engineering and architecture. Still, there have been some investigations of bamboo reinforcement related to seawater, though a very small number [11]. From an investigation in which concrete was mixed with seawater and its properties compared with those of concrete mixed with freshwater, electrolyzed water, and recycled water. This confirmed that the influence of mixing water on slump and strength is small. In spite of this, there remain some issues related to the use of bamboo reinforcement in seawater concrete. One is the issue of fungus on the bamboo reinforcement. Further, securing proper adhesion between concrete and bamboo is difficult and many different views exist regarding the basic properties of bamboo owing to the limited amount of research. Therefore, further research on bamboo reinforcement in seawater concrete is required, particularly relating to actual use in construction.

(4) FRP Reinforcement

Like bamboo, FRP is a material with very little susceptibility to corrosive deterioration. According to JSCE guidelines [19], concrete cracking resulting from FRP deterioration is negligible, which differentiates FRP from steel reinforcement. However, the guidelines call for surveys of crack formation in concrete reinforced with FRP materials from the viewpoint of aesthetic appearance and water tightness, and they specify an allowable crack width (e.g., the allowable crack width should be less than 0.5 mm in generally visible locations).

There are several forms of FRP, including rods, ropes, strands, grids, and sheets. There is an increasing number of instances of application of continuous fiber FRP sheets to strengthen concrete structures against earthquakes. In this section, the focus is on rod FRP from the viewpoint of replacing the steel reinforcement in concrete. There are currently very few reports related to seawater concrete. The price of FRP rods being relatively high compared with steel, evaluation of the economic efficiency of FRP using life cycle costing (LCC) is important. As a result, there are numerous studies related to the LCC efficiency of FRP rods [20]. Nakai et al. reported on exposure tests of concrete containing FRP rods and Pre-stressing steel in a marine environment (splash zone) and land-based environment. The results confirm that FRP rods are more durable than PC steel [22]. Nishimura et al. reported on the tensile strength characteristics of FRP rods (using aramid, glass, and carbon fibers) as determined in tests involving exposure to several kinds of marine environment [21]. They found the strength of FRP rods exposed on land to be lower than those exposed in a marine environment. Based on their results, it is considered that the main factor in strength degradation is not chloride ions but ultraviolet radiation. Therefore, it is considered that the use of FRP rods as reinforcement in seawater concrete is efficient.

BIBLIOGRAPHY

1. Yoshio Egusa, Hideki Watanabe, Takayuki Tamai, Naoto Jibiki, Kazuo Taka and Hideo Taka. 2005. Survey of bamboo reinforcement using electromagnetic radar method. Proceedings of Annual Conference of Architecture, (Kinki):697–698 (in Japanese).
2. Mariko Yamaoka, Masakazu Terai and Kouichi Minami Fundamental study of bending property of bamboo-reinforced concrete. Proceedings of Annual Conference of Architecture, (Hokuriku):235–236 (in Japanese).
3. Eiji Matsuo and Katsuhiko Takaumi. 2009. Application of bamboo for reinforcement of concrete. *Journal of JSCE F*, Vol.65, No.2:190–195 (in Japanese).
4. Seiya Shimoda, Ryuya Matsunaga, Sei Murakami, Yoshinori Kadono and Kouji Takeda. 2010. Experimental study of structural performance of bamboo concrete slab. Proceedings of Annual Conference of Architecture (Hokuriku):877–878 (in Japanese).
5. Hiroetsu Kikuchi and Katsunori Demura. 2008. Influence of curing condition on mechanical properties of mortar with bamboo reinforcement. *Proceedings of Japan Concrete Institute*, Vol.30, No.1:387–392 (in Japanese).
6. Masakazu Terai and Kouichi Minami. 2010. Fundamental study on adhesion and bending properties of bamboo-reinforced concrete. *Proceedings of Japan Concrete Institute*, Vol.32, No.2:1183–1188 (in Japanese).
7. Masakazu Terai and Koichi Minami. 2011. Fundamental study on compressive behavior at center part of bamboo-reinforced pier. *Proceedings of Japan Concrete Institute*, Vol.33, No.2:1171–1176 (in Japanese).
8. Kazuki Matsuda, Haruna Umemoto, Takuya Ikegawa, Hiroaki Kito, Hisao Kadokake and Hajime Oouchi. 2012. Fundamental study of recycled bamboo-reinforced concrete. Proceedings of Annual Conference of JSCE:1099–1100 (in Japanese).

9. Norio Endo, Takenori Inoue and Housei Matsuoka. 2007. Effect of bamboo on strengthening porous concrete. *Proceedings of Japan Concrete Institute*, Vol.29, No.2:319–324 (in Japanese).

10. Norio Endo, Hiroyuki Inose and Housei Matsuoka. 2010. Analysis of bending behavior of porous concrete reinforced by bamboo. *Proceedings of Japan Concrete Institute*, Vol.32, No.1:1379–1384 (in Japanese).

11. Katsuaki Horii. 2010. Experimental investigation of manufacturing of re-cycled waste concrete with bamboo reinforcement. Proceedings of Annual Conference of JSCE, Vol.61, V-457:909–910 (in Japanese).

12. Reiji Kawai, Masashi Kawamura and Yoshio Kasai. 2000. Adhesion behavior of soil cement reinforced with bamboo. *Proceedings of Japan Concrete Institute*, Vol.11, No.2:29–37 (in Japanese).

13. Katsuaki Horii, Norio Kurimeshihara, Shiho Hashimoto and Takashi Tada. 2007. Study on strength development of concrete made with waste materials and using bamboo reinforcement. *Proceedings of Japan Concrete Institute*, Vol.29, No.2:481–486 (in Japanese).

14. Horoetsu Kikuchi and Katsunori Demura. 2007. Mechanical properties of bamboo-reinforced cement mortar. *Proceedings of Japan Concrete Institute*, Vol.29, No.2:793–798 (in Japanese).

15. Shogo Kimura, Toshikatsu Saito and Katsunori Demura. 2012. Bending behavior of bamboo-reinforced cement mortar with emulsion-treated bamboo reinforcement. *Proceedings of Japan Concrete Institute*, Vol.34, No.1:1456–1461 (in Japanese).

16. Satoshi Nonoyama, Ayumu Ito and Keiichi Imamoto. 2011. Development of CFB using recycled plaster powder- fly ash or blast furnace slag- fly ash cement concrete and bamboo. *Proceedings of Japan Concrete Institute*, Vol.33, No.1:1505–1510 (in Japanese).

17. Masakazu Terai and Koichi Minami. 2012. Experimental study on mechanical properties of bamboo-reinforced concrete. *Proceedings of Japan Concrete Institute*, Vol.34, No.2:1279–1284 (in Japanese).

18. Kyou Kawamura. 1935. Bamboo reinforced concrete, Sankaido (in Japanese).

19. JSCE. 1998. Guideline for design and construction of concrete structures using continuous fiber reinforcement. Concrete Library 88 (in Japanese).

20. Hirofumi Watabe, Nobuyuki Nishikawa, Yuji Nakai and Kousuke Furuichi. 2003. Investigation of life-cycle cost of concrete with FRP. *Proceedings of Japan Concrete Institute*, Vol.25, No.2:1963–1968 (in Japanese).

21. Tsugio Nishimura, Taketo Uomoto, Yoshitaka Kato and Futoshi Katsuki. 1996. Tensile strength property of various kinds of FRP exposed to different environmental conditions. *Proceedings of Japan Concrete Institute*, Vol.18, No.1:1179–1184 (in Japanese).

22. Yuji Nakai, Hiroshi Sakai, Tsugio Nishimura and Taketo Uomoto. 2003. Mid-term report of exposure test of PC beam with various kinds of continuous fiber reinforcement. *Proceedings of Japan Concrete Institute*, Vol.25, No.1:335–340 (in Japanese).

3.3.4 Electric Current Control (Cathodic Protection, etc.)

Cathodic protection is widely used to protect steel structures and steel ships, etc. Its application to reinforced concrete (RC) or prestressed concrete (PC) is still unusual. However, in the recent few decades, this corrosion-protection technique has increasingly been applied to RC or PC structures as a repair method. Naturally, its application to seawater concrete from the initial stage is considered to be a good method to prevent corrosion of steel embedded in the concrete. A basic experiment is underway at Kyushu University yet, although it has been going for only a few years, the data already demonstrates its effectiveness [1, 2].

BIBLIOGRAPHY

1. Sabrina Harahap. 2019. *A study on corrosion behaviour of steel reinforcement in seawater-mixed concrete and application of corrosion prevention: Corrosion inhibitor and cathodic prevention.* Master's Thesis of Graduate School of Kyushu University.
2. Shunsuke Otani, Koji Yoshida, Sabrina Harahap, Daisuke Yamamoto and Hidenori Hamada. 2020. Research on application of cathodic protection to seawater-mixed concrete structures – test results up to 250 days –. Bosei-kanri (in Japanese).

3.3.5 Surface Coating

Surface coating is a technology normally used for inhibiting chloride ingress into concrete to protect it from chloride attack, especially under marine conditions. Surface coating can also prevent oxygen ingress into the concrete if the membrane is thick enough, meaning of the order of 1 mm in thickness. The corrosion process comprises an anodic reaction and a cathodic reaction. If one of the two reactions can be completely inhibited, the corrosion process will be stopped. Surface coating of the concrete with a reliable surface coating membrane achieves this by disturbing the oxygen supply to the cathodic area, halting corrosion even at high chloride concentrations such as in seawater concrete.

Research by Takeshi Oshiro and Shin Tanikawa is a good example of this concept, demonstrating that surface coating is a good method to protect steel bars in seawater concrete. Their experimental work is summarized in a book edited by R. Narayan Swamy and Shin Tanikawa [1].

BIBLIOGRAPHY

1. Narayan Swamy and Shin Tanikawa (editors); Hidenori Hamada, Jaw Chang Laiw and Takeshi Oshiro (co-authors). 2012. *Surface Coating for Sustainable Protection and Rehabilitation of Concrete Structures*. Japan: Kohbunsha Publishing Co., Ltd.

3.4 PERFORMANCE-BASED DESIGN FOR REINFORCED SEAWATER CONCRETE IN MARINE ENVIRONMENT

In this section, the basic concept of performance-based design is explained and the concepts of initiation and propagation and the lifetime of reinforced seawater concrete are introduced. Also, the use of BFS is calculated to be advantageous.

The contents are summarized as follows; our method for predicting initiation period, propagation period, and lifetime is shown. From our calculation results, propagation period and lifetime (initiation + propagation period) are almost the same for the mixed with freshwater and seawater. Especially, comparing '0.5 OPC' with freshwater and '0.5 B55' with seawater, the latter is much better than the former under the same water to cement ratio.

3.4.1 Outline

The outlook for modern standards and specifications in the 21st century is that they are going to be 'performance based.' This means that if lifetime serviceability and limit states can be assured for a specified number of years, then concrete structures could be designed with seawater in the mix. The question then is what kind of limit state requirements will there be over the lifetime of the structure. In performance-based design, for example, the following four serviceability limit states might be considered in relation to corrosion of steel reinforcement:

1) No corrosion at all
2) Corrosion permissible but no crack formation
3) Cracks permissible but no delamination
4) Delamination permissible as long as load-carrying capacity is sufficient

Here, we discuss only limit states (1) and (2).

First, we define the initiation period and propagation period as in Figure 3.3. Namely, the initiation period is defined as the time that elapses between construction and the start of corrosion, while the propagation period is defined as the time from corrosion starting to cracks appearing on the concrete surface due to corrosion. Further, lifetime is defined as the period from construction to the appearance of cracks, or the sum of the initiation period and the propagation period.

3.4.2 Calculation of Initiation Period

To calculate this period, at least four parameters are necessary: (1) Cover thickness, x, measured or from a drawing (and here assumed

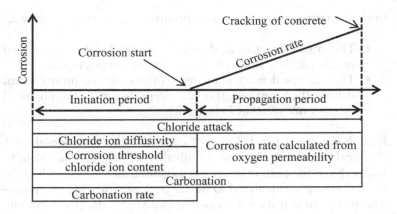

Figure 3.3 Outline of serviceability limit states.

to be 7 cm); (2) Surface chloride content, C_0, measured or taken to be 14 kg/m^3; (3) Diffusion coefficient of chloride ions in concrete, D_c, measured or a recommended value; and (4) Corrosion threshold chloride ion content, $C(x,t)$, a measured value. Using these values, the diffusion equation (from Fick's law) shown below is used to calculate the initiation period.

$$C(x,t) = C_0 \left(1 - \mathrm{erf} \left(\frac{x}{2\sqrt{D_c \cdot t}} \right) \right)$$

Namely, given x, C_0, D_c, and $C(x,t)$ and with some calculations, t = initiation period can be obtained.

3.4.3 Calculation of Propagation Period

To calculate this period, at least three parameters are necessary: (1) Cover thickness, x (here assumed to be 7 cm); (2) Corrosion rate or corrosion current density, measured or assumed; and (3) Critical amount of corrosion product on the surface of steel bar for crack propagation; assumed or obtained by a suitable equation (here, an equation proposed by Yokozeki et al.).

(1) Example of Corrosion Rate (Corrosion Current Density) Measurement

In this example, two major assumptions are necessary as follows:

- The corrosion current density is controlled by the oxygen supply and is assumed to be a constant corrosion rate.
- The current density is related to the cathodic limit current density (I_{lim}) (or the cathodic current density at the potential of −860 mV (Ag/AgCl electrode)).

From basic chemistry, 1 A.sec = 1 coulomb, 96,500 coulombs = 1 chemical eq., and one chemical equivalent of Fe_2O_3 is equivalent to 26.7 g. So, if the current density is measured as 1 μA/cm^2, then the corrosion rate is calculated to be 8.7 mg/year/cm^2. That is, if corrosion current density z is measured in μA/cm^2, the corrosion rate is 8.7z mg/year/cm^2.

(2) Critical Amount of Corrosion Product on Steel Bar: The Yokozeki Equation

The Yokozeki equation is as follows.

$$Wcr = -1.841\phi(\phi - 8.6661) + 145.1\alpha^{-1.194}$$

$$+ 3809A^{-0.8351} + 10.60X - 72.30$$

where Wcr is the amount of corrosion product per surface area of the steel bar at the end of propagation period (mg/cm^2), φ is the creep coefficient, α is the coefficient of thermal expansion (=3.2), A is the corrosion angle (360°) and X is a shape function (cover thickness/diameter of steel bar).

(3) Calculation of Propagation Period

The propagation period is calculated (in years) as *Wcr* divided by 8.7z, or the critical amount of corrosion product divided by the corrosion rate. If 1 uA/cm^2 is assumed to be the corrosion current density, the propagation period is $53.8/8.7 \fallingdotseq 6.2$ years.

3.4.4 Calculation Example (OPC versus OPC+BFS)

In this section, an example calculation is carried out. The required experimental values are obtained from mortar and concrete specimens. Other assumptions are:

- Environmental conditions: marine environment
- Cover thickness: 7 cm

(I) Outline of Specimens

Prism shape (40 × 40 × 160 mm) mortar specimens were used to obtain experimental values. As cement materials, OPC and OPC with BFS at 40%, 55%, and 70% replacement ratios were used as binders. And the water-to-binder ratios were 50% and 70%. Round steel reinforcing bars (yield point greater than 235 N/mm^2) were embedded in the mortar specimens with 10 mm cover thickness.

Moist curing was carried out for concrete specimens and a week of immersion curing for mortar specimens, respectively. The moist curing periods were five days for OPC concrete specimens and seven days for OPC with BFS concrete specimens. Immersion water was the same as the mixing water for each specimen. After curing, specimens were coated with epoxy resin before exposure except on the casting under-surface, which acted as the exposure surface.

(2) Measurements

For lifetime prediction (initiation period and propagation period), chloride ion diffusivity and the threshold chloride ion content for corrosion are needed to calculate the initiation period. For the propagation period, the steel corrosion rate (and also the cathodic limit current density (I_{lim}) from the assumption of oxygen supply control) and the conditions for cracking are needed.

- Chloride ion diffusivity: Chloride ion diffusivity was determined from the distribution of total chloride content at different depths in each specimen from the exposure side after four months of exposure.
- Corrosion current density: The polarization resistance was measured by the AC impedance method using high frequency (10 kHz) and low frequency (10 mHz) alternating current and corrosion current density was calculated with the Stern-Geary constant [1] based on the following equation:

$$I_{corr} = \frac{K}{R_{ct} \cdot S}$$

where I_{corr} is the corrosion current density in $\mu A/cm^2$, R_{ct} is the polarization resistance in Ω, S is the surface area of steel bar (40.82 cm^2), and K is the Stern-Geary constant (0.0209 V).

- Corrosion threshold chloride ion content: the threshold chloride ion content was measured as the chloride ion content in the mortar around the steel bar when the corrosion current density reached 0.2 $\mu A/cm^2$ [2]).
- Limit current from the assumption of oxygen supply control: I_{lim} was obtained from electro-chemical measurements using cathodic polarization curves as shown in Figure 3.4.

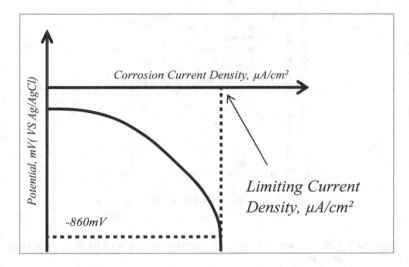

Figure 3.4 Concept of limiting current density measurement.

BIBLIOGRAPHY

1. Stern, M. and Geary, A. L. 1957. Electrochemical polarization: I. A theoretical analysis of the shape of polarization curves. *Journal of the Electrochemical Society*, Vol.104, No.1:56–63.
2. CEB Working Party V/4.1. 1997. Strategies for testing and assessment of concrete structure affected by reinforcement corrosion (draft 4). BBRI-CSTC-WTCB.

(3) Results

- Chloride ion diffusivity: Figure 3.5 compares the chloride ion diffusivity of concretes with BFS mixed with seawater and freshwater. The chloride ion diffusivity of concrete mixed with seawater was lower than that with freshwater. This might be due to the presence of initial chloride content induced by the seawater, which could reduce the chloride concentration gradient between the concrete and surrounding seawater. (In Figures 3.5 to 3.8, '0.5OPC' means concrete with W/B = 0.5 and OPC binder, '0.5B40' means concrete with W/B = 0.5 and a BFS replacement ratio of 40%, etc. Also, the horizontal axis represents freshwater and the vertical axis seawater.)

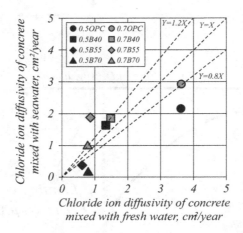

Figure 3.5 Chloride ion diffusivity of concrete mixed with freshwater versus seawater.

- Corrosion threshold chloride ion content: In this example, the value is 53.8 mg/cm^2 for the case of 70 mm cover depth and 13 mm diameter steel reinforcing bar.

(4) Estimation of Initial and Propagation Periods and Lifetime

- Estimation of initiation period: Figure 3.6(a) compares the initiation period of concretes with BFS mixed with seawater and freshwater. The initiation period in the case of seawater is shorter than that with freshwater. It is considered that the seawater in the mix induces an initial chloride content, reducing the time taken to reach the corrosion threshold. This means that, if initiation period only is considered, BFS concrete mixed with seawater is not suitable for structures with respect to reinforcement corrosion.

  Figure 3.6(b) compares the propagation period of concretes with BFS mixed with seawater and freshwater. The propagation period of concrete mixed with seawater is longer than that with freshwater.
- Lifetime (initiation + propagation period): Using the above results, the lifetime is calculated as the summation of the

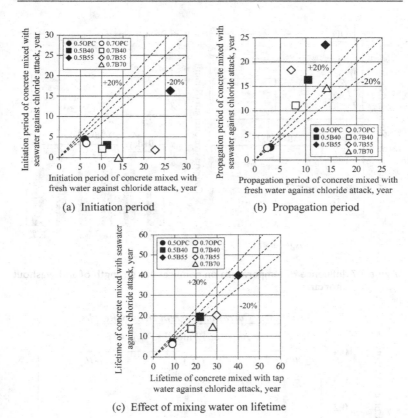

(a) Initiation period

(b) Propagation period

(c) Effect of mixing water on lifetime

Figure 3.6 Effect of mixing water on the initiation period, the propagation period and lifetime.

initiation period and the propagation period. The results are shown in Figure 3.6(c). Although when the initiation period only is considered, concrete with BFS mixed with seawater may not be suitable for use in construction. However, when the summed initiation and propagation periods are taken as the concrete lifetime, the seawater concrete reaches almost the same or a slightly shorter lifetime than that mixed with freshwater. In the case of W/B = 0.5 and a BFS replacement ratio of 55% in particular, the lifetime difference between seawater and freshwater concrete is a minimum. The lifetime of this concrete mixed with seawater is about 40 years.

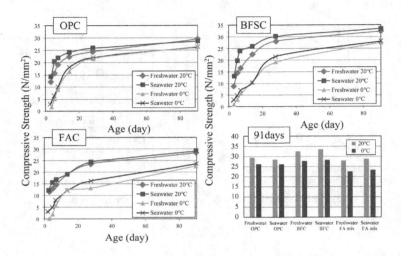

Figure 3.7 Influence of temperature on compressive strength of anti-washout mortar.

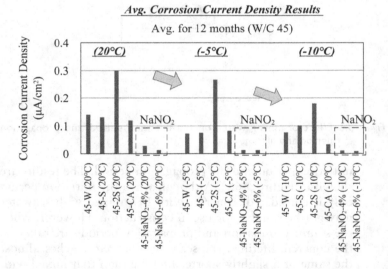

Figure 3.8 Influence of temperature and mixing water on the corrosion current density of steel reinforcing bars in mortar (W: mixed with distilled water; S: mixed with seawater; 2S; mixed with two times seawater chloride content; CF: mixed with seawater and calcium aluminate; $NaNO_2$: mixed with seawater and $NaNO_2$) (Calcium aluminate and $NaNO_2$ are expected to protect the concrete from freezing as well as mitigating corrosion).

(5) Conclusion

Taking a performance-based design approach, the lifetime (initiation period + propagation period) of concrete mixed with seawater is not very different from that mixed with freshwater. In particular, comparing 0.5OPC mixed with freshwater against 0.5B55 mixed with seawater, the seawater concrete is much better at the same W/B.

BIBLIOGRAPHY

1. Otsuki, N., Nishida, T., Yi, C., Nagata, T. and Ohara, H.. 2014. Effect of blast furnace slag powder and fly ash on durability of concrete mixed with seawater. 4th International Conference on the Durability of Concrete Structures.
2. Yokozeki, K., Motohashi, K., Okada, K. and Tsutsumi, T. 1997. A rational model to predict service life of RC structures in marine environment. Fourth CANMET/ACI International Conference on Durability of Concrete:777–796.

3.5 THE EVOLUTION OF DESIGN METHODOLOGY

The specification-based design has been the norm for concrete structures, but this is now shifting performance-based design is coming to the fore. The use of seawater (as mixing or curing water) can now be considered and investigated from the viewpoint of performance-based design. The basic concept of performance-based design as outlined by JSCE and FIB is explained below.

The JSCE [1] provides a definition in the JSCE standards: 'The basics of performance verification.' In cases when quantitative verification of performance-based design is needed, the principle procedures are as follows. Firstly, appropriate parameters for the designated performance and the limit values of the parameters should be determined. Secondly, the performance of the structure during its construction and design service life should be shown to satisfy the designated performance by numerical analysis, testing, and/or calculation of response by a reliable method. The basic concept of performance-based design given by FIB [2] is in section 3.5.1 of the FIB Model Code for Concrete Structures 2010.

3.5.1 FIB Model Code

The performance of a structure or a structural component refers to its behavior consequent to actions to which it is subjected or which it generates. Structures and structural members must be designed, constructed, and maintained in such a way that they perform adequately and in an economically reasonable way during construction, service life, and dismantlement.

In general: (1) Structures and structural members must remain fit for the use for which they have been designed, (2) Structures and structural members must withstand extreme and/or frequently repeated actions and environmental influences liable to occur during their construction and anticipated use, and must not be damaged by accidental and/or exceptional events to an extent that is disproportional to the triggering event, and (3) Structures and structural members must be able to contribute positively to the needs of humankind with regard to nature, society, economy, and well-being.

Accordingly, three categories of performance have to be addressed:

1) Serviceability, meaning the ability of a structure or structural members to perform adequately, with appropriate levels of reliability, in normal use under all (combinations of) actions expected during its service life.
2) Structural safety, meaning the ability of a structure and its members to guarantee overall stability, adequate deformability, and ultimate bearing resistance corresponding to the assumed actions (both extreme and/or frequently repeated actions and accidental and/or exceptional events) with appropriate levels of reliability for the specified reference periods. Structural safety must be analyzed for all possible damage states and exposure events relevant to the design situation under consideration.
3) Sustainability, meaning the ability of a material, structure, or structural members to contribute positively to the fulfillment of the present needs of humankind with respect to nature and human society, without compromising the ability of future generations to meet their needs in a similar manner [2].

Namely, in these performance-based design methodologies, it is emphasized that 'in principle, it should be verified that the structures and structural members must remain fit for the use for which they have been designed during construction and the designed service life.' The use of seawater could be considered if these principles are taken into account. It should be not so difficult to design concrete members to comply with the principles during their expected service life, as follows:

1) Regarding strength, ensure that a certain compressive strength (ON/mm^2) is maintained.
2) Regarding durability, ensure that cracking as a result of steel corrosion does not occur.
3) Regarding load-carrying capacity, ensure that load-carrying capacity remains above 80% of the initial capacity.

Further, there could be many variations. Relating to (2) in particular, it is possible to use ordinary steel reinforcement when the expected service life is short and to use the cathodic protection method, stainless steel, and/or coated steel depending on the technological viewpoint. Of course, cost-benefit analysis is also necessary.

As noted in an earlier chapter, almost all standards and recommendations in the world today proscribe or limit the use of seawater as mixing and/or curing water for concrete. However, this chapter has shown that taking into account the concept of performance-based design, seawater could be used as mixing and/or curing water for the construction of concrete structures with proper verification.

BIBLIOGRAPHY

1. JSCE. 2012. Standard Specification for Design and Construction of Concrete Structures (General Principles), 20.
2. FIB. 2013. fib Model Code for Concrete Structures 2010, 21.

3.6 MINIMUM COVER THICKNESS

In performance-based design, the cover thickness should be designed based on performance verification. However, from long experience with many structures it is recommended that the cover thickness should be at least 50 mm, and if possible 70 mm, for normal concrete structures, which limits oxygen diffusion into the concrete to the minimum possible to prevent steel corrosion progress.

3.7 SEAWATER CONCRETE IN COLD CLIMATES

3.7.1 Introduction

In this subsection, the performance of seawater concrete in cold climates is discussed, mainly based on Dr. Aung Minn's doctoral thesis [1] and two other papers [2, 3]. In this 21st century, many possibilities are opening up to develop regions with cold climates, including the arctic regions. Arctic regions are known to have little rainfall and a lack of freshwater, so the use of seawater concrete would be very advantageous. Further, aside from not wasting freshwater, other important advantages of seawater in concrete are anti-freezing and quick hardening properties and a lower corrosion rate in cold temperatures. Also, in this subsection, the use of seawater in anti-washout concrete (mortar) is introduced. Finally, it can be confirmed that seawater-mixed mortar performed well under cold temperature conditions in terms of strength, porosity, corrosion rate, and chloride diffusivity, and there is almost no difference mixed with freshwater or seawater.

3.7.2 Compressive Strength in a Cold Climate

(1) Cast and Cured in the Atmosphere (−5°C and −10°C Cases)

The influence of temperature on the compressive strength of mortar was tested. In the test reported in this work, cylindrical mortar specimens measuring φ 50 mm × 100 mm were used. Specimens were cast with OPC and two types of mixing water (freshwater and artificial seawater). Under a curing temperature of −5°C, the seawater-mixed specimens increased strength continuously until 91 days. In experimental series 2S (with twice the chloride content of seawater), CF specimens (using $CaO \cdot 2Al_2O_3$) as a chloride fixing admixture and with the addition of $NaNO_2$ (4%) show better strength development rate. On the other hand, freshwater mixed specimens show only slight strength development until 7 days and almost no further strength development beyond 7 days. It seems the hydration process in freshwater mixed specimens is retarded due to the low temperature. This is evidence that the concrete froze in the case of the freshwater mixed specimens. The following reasons for the seawater concrete having better strength development up to 91 days at −5°C may be considered:

1) Various salts in seawater accelerate the cement hydration process.
2) Seawater shorten the setting time of cement and protects the concrete from freezing for further strength development.
3) Seawater reduces the freezing point of the mixing water contained in the mortar and maintains it in the liquid phase, allowing cement hydration to continue.

The results also confirm that the corrosion remediation admixture, sodium nitrite ($NaNO_2$), and chloride fixing admixture with seawater gave even better strength development than the case only with seawater. The eutectic temperature of $NaNO_2$ is −19.7°C and it further reduces the freezing point of seawater and maintains it in the liquid phase for cement hydration even at curing temperatures under −5°C. On the other hand, the chloride fixing admixture ($CaO \cdot 2Al_2O_3$) is one kind of calcium aluminate. Calcium aluminate cement has a property of significantly higher early strength gain up to 39 N/mm^2 at 0°C and it can be placed in very cold weather. Hence, 10% replacement of cement with chloride fixing additive can achieve early strength development and protect the concrete from freezing even at a curing temperature of −5°C. So although sodium nitrite ($NaNO_2$) and chloride fixing additives ($CaO \cdot 2Al_2O_3$) are added to seawater concrete for the purpose of corrosion remediation, these admixtures also contribute to the material properties of seawater concrete in cold weather conditions to enhance strength development.

At a curing temperature of −10°C, neither seawater nor freshwater concrete gain strength until 91 days. This means that without special antifreeze admixture or additional heating, both seawater and freshwater concrete will fail to gain the required strength at such temperatures.

To understand the ultimate strength development of seawater concrete compared with that of freshwater concrete, the 91-day strengths under several different curing conditions in all experiments are compared. The results confirm that the seawater concrete achieves a minimum 91-day strength of 30 N/mm^2 under $-5°$C curing and achieved up to 50 N/mm^2 in the cases of 2S, CF, and NaNO$_2$ (4%) for W/C 45%, while freshwater concrete achieved less than 20 N/mm^2 at the same curing temperature.

It is therefore confirmed that seawater concrete exhibits better strength development than freshwater concrete. All cases of seawater concrete gain strength continuously up to 91 days and reach a compressive strength of more than 30 N/mm^2. Seawater concrete that includes the corrosion remediation admixture sodium nitrite and a chloride fixing additive exhibits better strength development than plain seawater concrete at $-5°$C. However, neither seawater concrete nor freshwater concrete gain strength when cured at $-10°$C.

(2) Cast and Cured in Seawater
(Anti-washout Mortar)

The influence of temperature on the compressive strength of anti-washout mortar is shown in Figure 3.7. Anti-washout mortar (concrete) is expected to be cast and cured in seawater between a moderate temperature of 20°C and a cold temperature of 4°C, so the tests described in this report were at 20°C and 0°C. It is confirmed that compressive strength increases with time even in 0°C specimen, which indicates that seawater-mixed anti-washout concrete performs well in a cold environment. No evidence of concrete freezing was found. At an early age, the seawater-mixed specimens show greater compressive strength than specimens mixed with freshwater. However, this difference in compressive strength by type of mixing water (freshwater versus seawater) has disappeared by the age of 91 days.

The compressive strength tests were on cylindrical mortar specimens measuring φ 50 × 100 mm. Two types of mixing water (freshwater and seawater) and three types of cement (OPC, BFS cement, and FA cement) were used to make the specimens. Specimens were cured under submerged conditions at two water temperatures (0°C and 20°C). The specimens were cast with the mold submerged in artificial seawater 10 cm below the surface. A spoon was used to pour the mortar into the mold and no tamping or vibration was used. Specimens were cured in the artificial seawater until the test age. The temperature was controlled during the casting and curing of the specimens. Compressive strength tests were conducted at 3, 5, 7, 14, 28, and 91 days of age.

3.7.3 Corrosion Rate (Corrosion Current Density) in a Cold Climate

(1) Cast and Cured in the Atmosphere (−5°C and −10°C Cases)

The influence of temperature on the current density of steel reinforcing bars in mortar is shown in Figure 3.8.

These measurements were carried out using mortar specimens measuring φ 9 mm × 100 mm with SR235 reinforcement (a round steel bar with a yield point over 235 MPa) to investigate the corrosion behavior of steel in seawater concrete. The steel bar was placed centrally with a cover thickness of 10 mm. Note that a current density of 0.2 µA/cm^2 is taken to be the threshold current density at which steel corrosion is initiated inside the concrete [4].

From Figure 3.8, it is clear that all current densities, except case 45-2S (W/C = 0.45 with double seawater chloride content), remain below 0.2 µA/cm^2. Also, the lower the temperature, the lower the measured current density. In particular, cases with NaNO$_2$ exhibit current densities less than 0.05 µA/cm^2. This means that, from an engineering viewpoint, corrosion can be neglected for at least 100 years.

(2) Cast and Cured in Seawater (Anti-washout Mortar)

The influence of temperature on the current density of steel bars in mortar exposed to seawater for five months was tested. In this case, mortar specimens measuring φ 9 mm × 100 mm with SR235 reinforcement were used for steel bar corrosion behavior measurements and oxygen permeability calculations. Again, the steel bar was centrally placed and the cover thickness was 10 mm.

Test results indicate that temperature (0°C versus 20°C) has more influence than the type of mixing water (seawater versus freshwater). BFS cement and FA cement mixes exhibit higher corrosion current densities than OPC. This may be due to the lower alkalinity of BFS and FA cements compared to the OPC mix. Also, from an engineering viewpoint, the corrosion rate is negligible (no corrosion cracking not expected 100 years).

3.7.4 Consistency of Anti-washout Mortar

The consistency flow test of each experiment case was carried out. In this case, consistency flow tests are conducted (in accordance with the Japanese architectural standard specification) by using OPC mortar with water contents of 350, 365, and 380 kg/m³. In order to control the test temperature at 0°C and 20°C, the flow tests were implemented under water. It was found that both seawater usage and a cold environment contribute to reduced consistency. However, the same consistency value as with freshwater at 20°C can be achieved in seawater at 0°C by increasing the water content by 24 kg/m³. In terms of consistency, seawater can be used in a cold environment with the appropriate design of mix proportion.

BIBLIOGRAPHY

1. Aung Kyaw Min. 2017. *Influence of cold temperature on steel corrosion in seawater mixed concrete exposed to marine environment.* PhD diss., Tokyo Tech.
2. Aung Kyaw Min, Mitsuyasu Iwanami, Nobuaki Otsuki, Keisuke Matsukawa and Yuma Yoshida. 2016. Study on construction performance and durability of seawater mixed anti-washout concrete in cold environment. 41st Conference on Our World in Concrete & Structures:24–26.
3. Aung Kyaw Min, Mitsuyasu Iwanami, Nobuaki Otsuki and Keisuke Matsukawa. 2017. Durability of seawater mixed concrete against rebar corrosion in cold temperature condition. ACI Fall 2017, Anaheim (Research in Progress session).
4. CEB (European Committee for Concrete) and Belgian Building Research Institute. 1997. *Strategies for Testing and Assessment of Concrete Structure Affected by Reinforcement Corrosion,* International Federation for Structural Concrete, Lausanne, Switzerland.

3.8 SEAWATER CONCRETE WITH SHIRASU

3.8.1 Shirasu

(1) Volcanic Sediment as a Construction Material

Volcanic sediment has long been used as a construction material and was known even in the Roman era. Structures built using them are still extant at present. In Japan, it is around 120 years since volcanic sediment was first used as a construction material. In the Meiji era (1868–1912), concrete mixed with volcanic ash and limestone was used in various structures. The technology used in that construction work was mainly imported from western countries. The most known work of that time is the breakwater at Otaru port, located in central Hokkaido in the far north of Japan. The work's director, Dr. Hiroi, published a paper on concrete block production for the breakwater. His paper includes details of material quality, concrete mix proportion, manufacturing method, and curing method for the production of highly durable concrete. Results from the mortar briquettes that he tested explain the role of volcano ash and the importance of adequate consolidation. Even now tests are continuing using specimens that Dr. Hiroi made more than 100 years ago.

(2) Properties of Shirasu

Shirasu is a Japanese word for a particular kind of pyroclastic flow deposit found in Kagoshima, southern Kyushu. The word comes from an old dialect of Kagoshima. It has become recognized as a good pozzolanic material for concrete production. Normally, shirasu is used as a fine aggregate or as a mineral admixture (classified fine particles).

Shirasu differs from lava or volcano ash in that, in its natural state, each particle is firmly aggregated into a solid. However, once disturbed, shirasu becomes sand or silt. The disturbed shirasu contains volcanic glass of several micrometers as well as pumice stones up to several tens of centimeters in diameter, so it is a mixture of fine particles and lightweight stones. In Japan, it has become commonplace to use this kind of volcanic sediment in concrete production. In this section, we describe seawater concrete using volcanic sediment.

Figure 3.9(a) shows examples of the particle size distributions of various types of shirasu from the Kagoshima district [1]. Yoshida shirasu is sediment found in river beds and lakes; its particles are all less than 75 μm. Other shirasu are pyroclastic flow deposits. Ata shirasu comes from the Ata pyroclastic flow, which contains finer particles than the Yokogawa and Kushira shirasu obtained from the Ito pyroclastic flow. The Ata pyroclastic flow occurred 70,000 years earlier than the Ito pyroclastic flow, which explains its finer particle size. Figure 3.9(b) shows X-ray diffraction (XRD) patterns for each shirasu [1, 2]. These confirm that Yoshida shirasu

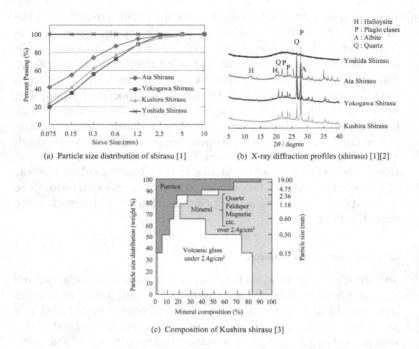

(a) Particle size distribution of shirasu [1]

(b) X-ray diffraction profiles (shirasu) [1][2]

(c) Composition of Kushira shirasu [3]

Figure 3.9 Particle size distribution, chemical and physical characteristics of shirasu [1–3].

is an amorphous volcanic glass. The others contain quartz (Q), albite (A), and plagioclases (P), among others. Ata shirasu contains a large amount of albite compared to Yokogawa shirasu or Kushira shirasu. All also contain Halloysite (H), a kind of clay mineral.

The mineral composition of Kushira shirasu has been categorized by Tomoyori [3], as shown in Figure 3.9(c). This shows the tendency for larger particles to include a lot of minerals such as quartz or plagioclases, while in finer particles there is a high content of volcanic glass. The results explain that shirasu is a pyroclastic flow deposit composed of minerals such as quartz and plagioclases, as well as pozzolans, such as volcano ash.

(3) Reactivity of Shirasu

Figure 3.10(a) shows the reaction rate of cement pastes with 20% replacement by shirasu, showing that shirasu reacts over the long

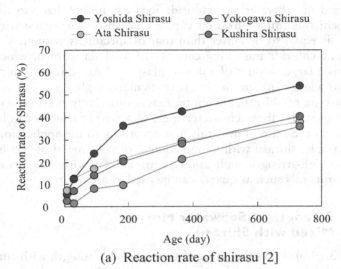

(a) Reaction rate of shirasu [2]

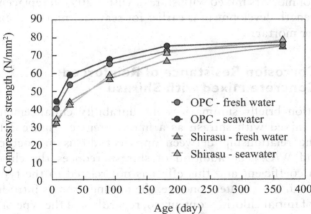

(b) Relationship between curing time and compressive strength

Figure 3.10 Reaction rate and compressive strength development of shirasu [2].

term. This means that the strength of concrete using shirasu increases over the long term. In particular, Yoshida shirasu, which is mainly composed of volcanic glass, shows a higher reaction rate, indicating that volcanic glass greatly affects the reaction rate of cement with shirasu. When a specimen made with shirasu is immersed in saltwater, the chloride ions are immobilized in the specimen; this immobilization capacity of specimens made with shirasu is equal to or better than that of specimens without shirasu. The chloride ion fixing capacity of Yoshida shirasu, which contains a large amount of volcanic glass, and Ata shirasu, which contains a large amount of clay, is particularly high.

Each type of shirasu, therefore, reacts differently with cement according to the above characteristics, so a type of shirasu can be selected by the ratio of minerals in it according to the application. For example, shirasu with a high content of volcanic glass can be used for high-strength applications, while shirasu with a high content of minerals such as quartz can be used as aggregate.

3.8.2 Strength of Seawater Mortar Mixed with Shirasu

Figure 3.10(b) shows the change in compressive strength with curing time of mortars mixed with shirasu with a 20% in replacement ratio. Strength development is earlier for seawater mortar than for freshwater mortar.

3.8.3 Corrosion Resistance of Reinforced Concrete Mixed with Shirasu

This section briefly summarizes the durability enhancement of concrete mixed with shirasu as a fine aggregate. Figure 3.11(a) shows the relationship between apparent diffusion coefficient (Dap) and W/C. The addition of shirasu reduces the chloride diffusion coefficient and this effect is not related to the type of cement used. In seawater concrete, the mixing process introduces $3 \ kg/m^3$ of initial chloride content, so, regardless of the type of fine aggregate, there is a corrosive environment around the reinforcing steel and the corrosion rate after corrosion initiation is a technical problem.

Figure 3.11(b) shows the relationship between initial chloride content and the corrosion rate of steel reinforcement. Here, the W/C of the concrete is 50% and the cover depth is 42 mm. These are the results of a four-year exposure test in a marine atmosphere and in the tidal zone. From this figure, it can be said that concrete with shirasu suffers a reduced steel corrosion rate. In fact, the corrosion rate with a $3 \ kg/m^3$ initial chloride content is negligible.

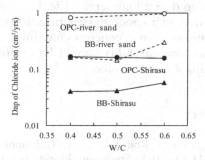

(a) Relationship between apparent diffusion coefficient and W/C [4]

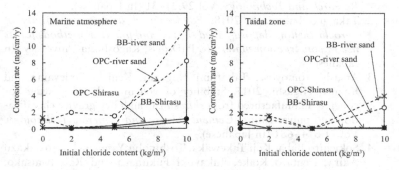

(b) Relationship between initial chloride content and corrosion rate of steel bars

Figure 3.11 **Apparent diffusion coefficient and corrosion of steel bars in shirasu mixed concrete [4].**

The ability of shirasu to reduce corrosion under higher chloride contents is also recognized.

Using these experimental results, the effectiveness of shirasu in seawater concrete can be explained as follows. First, in seawater concrete, the reinforcing steel is exposed to a corrosive environment, but the corrosion rate is lower in concrete containing shirasu. By evaluating the corrosion rate, the long-term durability can be established and, if necessary, countermeasures to enhance durability could be designed. On the other hand, in a chloride-rich environment, such as under marine conditions, chloride ions intrude from outside a structure during the service life. Shirasu has the ability to reduce the ingress of chloride ions from outside and also decrease the corrosion rate. Consequently, shirasu can be used to reduce corrosion over a long service life. However, corrosion of the reinforcing steel in a concrete structure depends on the environmental conditions, even in concrete containing shirasu. For this reason, evaluations of durability should be carefully undertaken before determining a concrete mix.

BIBLIOGRAPHY

1. Takayuki Fukunaga, Koji Takewaka, Toshinobu Yamaguchi and Yoshikazu Akira. 2018. Study on chloride penetration resistance of mortar using *shirasu* as cement replacement material. *Concrete Research and Technology*, Vol.29:21–31 (In Japanese).
2. Takayuki Fukunaga, 2018. *Fundamental study on reactivity of pyroclastic flow deposit called shirasu and practical methods for its use in construction materials*, PhD diss, Kagoshima University (In Japanese).
3. Atsushi Tomoyose, Takafumi Noguchi, Kenichi Sodeyama and Kazuro Higashi. 2017. Stability of volcanic silicate powder for concrete manufactured from *shirasu* through dry gravity classification and pulverization. *Cement Science and Concrete Technology*, Vol.71:674–681 (In Japanese).
4. Sokichiro Baba, Koji Takewaka, Toshinobu Yamaguchi, Yoshikazu Akira, Kentaro Koike, Takayuki Fukunaga and Ran Iwaisako. 2020. Experimental study on performance evaluation of *shirasu* concrete in chloride attack environment. Advances in Construction Materials, Proceedings of the ConMat'20, Vol.6:350–360.

Part II

Actual Constructions

Chapter 4

Japanese Experience with Seawater Concrete

4.1 INTRODUCTION

Most of the world's construction standards prohibit the use of seawater in concrete production for RC and PC. However, it is a fact that there are many existing structures made with seawater concrete, some dating back to more than five decades. In this chapter, four examples of the existing seawater concrete structures in Japan are introduced.

4.2 OLDER STRUCTURES OF SEAWATER CONCRETE

In this section, two examples of the existing structures made in the past with seawater concrete are introduced. One is an RC lighthouse and the other a breakwater constructed by the pre-packed concrete method. In both cases, blast furnace slag (BFS) cement was used. Both remain in good condition and are in service even at present, several decades after construction.

4.2.1 Lighthouse on Ukushima Island, Nagasaki Prefecture, Japan

There is a documented record [1] of three lighthouses made with seawater concrete in Japan (two in Nagasaki Prefecture and one in Yamaguchi Prefecture). Here, the Uku-Nagasakibana Lighthouse in Nagasaki prefecture is introduced as one example. This lighthouse is located on a small isolated island (Ukushima island) and was erected in a location with high waves. Freshwater could

DOI: 10.1201/9781003194163-4

not be used for concrete mixing, so seawater concrete was used for construction. The lighthouse was built between August and October 1959.

Cement, coarse aggregate, and fine aggregate were BFS cement type B (JIS R 5210 'Portland Cement'), gravel taken from the seacoast, and sand taken from the seacoast, respectively. The design strength of the concrete (σ_{28}) was 280 kgf/cm^2 (28 MPa). According to the record [1], (a) the mass ratio of cement to fine aggregate to coarse aggregate was 1:2:4, (b) the unit cement content was 340 kg/m^3, (c) the slump value was 15 cm, (d) W/C was 40–60%, and (e) 1% of calcium chloride ($CaCl_2$) was used in addition to seawater. Furthermore, for the rubble concrete, the rubble diameter and unit cement content were 120 mm and 140 kg/m^3, respectively.

Figure 4.1(a) shows a cross-sectional view of the lighthouse. Although construction was completed more than 60 years ago, no serious abrasion of the mortar layer or other damage has been found on the concrete surface, and the lighthouse remains in very good condition.

BIBLIOGRAPHY

1. Yawata Chemical Co. 1963. Portland blast furnace cement construction examples (civil engineering, water contact construction). Technical Data:81–85 (in Japanese).

4.2.2 Port Breakwater Constructed by Pre-packed Concrete Method

The results of an investigation of the physical properties of concrete samples taken from an actual port structure (a breakwater) in Tajiri Port, Japan, are outlined. The breakwater was made with seawater concrete and constructed using the pre-packed concrete method more than 50 years ago. The physical properties investigated were compressive strength and the distribution profile of chloride ions.

Tajiri Port is located on an indented coastline (the Rias coastline) in Tottori Prefecture. According to records of the mixing water used for the pre-packed concrete, seawater was used for sections of the structure built until 1963 and freshwater since 1967. The survey

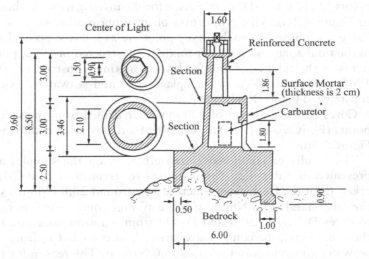

(a) Sectional view of Uku-Nagasakibana Lighthouse [1]
(Hatched areas indicate concrete mixed with seawater) (Unit: m)

(b) Uku-Nagasakibana Lighthouse (at present)

Figure 4.1 (a) Sectional view of Uku-Nagasakibana Lighthouse [1] (Hatched areas indicate concrete mixed with seawater) (Unit: m). (b) Uku-Nagasakibana Lighthouse (At present).

points (B), (C), and (D) were chosen for the investigation, as shown in Figure 4.2(a). Mix proportion of grouting mortar is given in Table 4.1. Outline of each survey point are as follows: point (B) is in the tidal zone and seawater mixing, constructed in 1962, point (C) is in the tidal zone and freshwater mixing, constructed after 1967, and point (D) is in the splash zone and seawater mixing, constructed before 1962.

Cores with a diameter of 100 mm were drilled from the survey points (B), (C), and (D). An example of a drilled core is shown in Figure 4.2(b).

The results of compressive strength tests on the samples are presented in Table 4.2. The compressive strength of core C-1(b) taken from the freshwater concrete was around 20.8 N/mm². On the other hand, 25.5 N/mm² was the average compressive strength of cores D-1(b), D-2(b), and D-3(b) from seawater concrete. On the other hand, the compressive strength of cores B-1(b) from the seawater concrete was 13.6 and 15.0 N/mm². The reason for the lower value is thought to be the presence of large coarse aggregate in the core relative to the diameter of the core.

The distribution of chloride ions (Cl⁻) in the mortar part of the cores is shown in Figure 4.3. Chloride ion content was measured in only the mortar after removing the coarse aggregate. Measurements of total Cl⁻ content and soluble Cl⁻ content were according to JIS A 1154 'Method of test for chloride ion content in hardened concrete'. The fixed Cl⁻ content is defined as the difference between the total and the soluble Cl⁻ values.

As seen in Figure 4.3(a) [survey point (B) seawater concrete], the Cl⁻ content peaks at a depth of 10 mm from the surface, where the total Cl⁻ content was approximately 20 kg/m³. At depths greater than 100 mm, the total Cl⁻ content was approximately 6 kg/m³. It can be inferred that Cl⁻ ions supplied by the mixing seawater have remained inside the concrete over the long term at survey point (B).

According to Figure 4.3(b) [survey point (C) freshwater concrete], the Cl⁻ content decreased rapidly with depth from the surface and at depth below 200 mm was approximately 1 kg/m³. The mortar at survey point (C) was mixed with freshwater, so the initial Cl⁻ content is assumed to have been very low. The survey point was in the tidal zone, so foreign chloride ions might have penetrated over the long period since construction.

(a) Survey point (B) (seawater concrete, tidal zone) and
Survey point (D) (seawater concrete, splash zone)

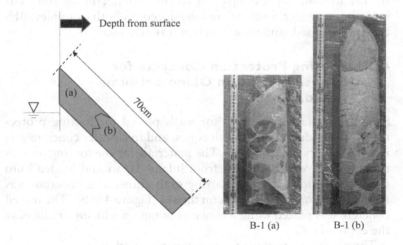

(b) Core sampling (B-1)

Figure 4.2 (a) Survey point (B) (seawater concrete, tidal zone) and Survey point (D) (seawater concrete, splash zone). (b) Core sampling (B-1).

Table 4.1 Mix proportion of grouting mortar (Tajiri Port) [1]

Construction year	Mix proportion of grouting mortar				Flow time (sec)	Remarks σ_{28} : Design value
	S/C	DA/C	Al/C	W/C		
1958	1.0	0.1%	0.01%	50%	15–19	σ_{28} = 125 kgf/cm^2 (12.5 MPa)
1959	1.0	0.1%	0.01%	55%	15–19	σ_{28} = 100 kgf/cm^2 (10 MPa)
1960	1.0	0.1%	0.01%	53%	15–19	σ_{28} = 100kgf/cm^2 (10 MPa)

C: Blast furnace slag cement (Yahata Steel Company; slag replacement ratio = 30%)

S: Uratomi beach sand (F.M.=1.11)

DA: Foaming agent

Al: Bloating (Foaming) agent (Aluminum powder)

W: Seawater

4.3 RECENT MANUFACTURING AND CASTING OF SEAWATER CONCRETE

In this section, two examples of recent manufacturing and casting using seawater concrete are introduced. In both examples, BFS cement was used, and in one, debris was also used.

4.3.1 Footing Protection Concrete for Exposed Rocks on Okinotorishima Island, Japan [1]

Seawater concrete was used for walls placed for footing protection in the undersea and splash zones, and freshwater concrete was used in the dry environment. The materials for the footing protection concrete were obtained from inland Japan and loaded onto a mother vessel (Figure 4.4(a)) by grab crane. The concrete was mixed on a working platform on the sea (Figure 4.4(b)). The mixed concrete was placed using a concrete pump on a lifting cradle near the exposed rock.

The general conditions for this concrete work were:

1) Air temperature = 30°C or more.
2) Anti-washout underwater concrete was used.
3) The pumping distance was relatively long.

Table 4.2 Results of compressive strength tests

No.	d (mm)	h (mm)	A coefficient (h/d)	Max. load (kN)	Compressive strength (N/mm²)	Ave. compressive strength (N/mm²)
B-1 (b) Upper Seawater	100.3	188.1	0.992	119.8	15.0	14.3
B-1 (b) Lower Seawater	100.4	186.1	0.990	108.9	13.6	
C-1 (b) Freshwater	100.3	199.2	0.998	164.8	20.8	20.8
D-1 (b) Seawater	100.3	200.2	0.999	185.7	23.5	
D-2 (b) Seawater	100.3	200.3	0.999	219.4	27.7	25.5
D-3 (b) Seawater	100.3	197.0	0.997	200.2	25.3	

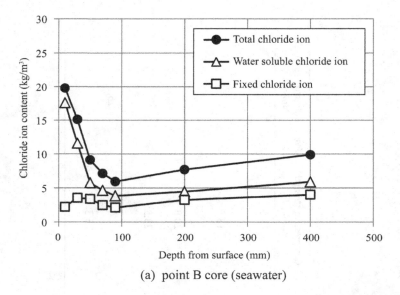

(a) point B core (seawater)

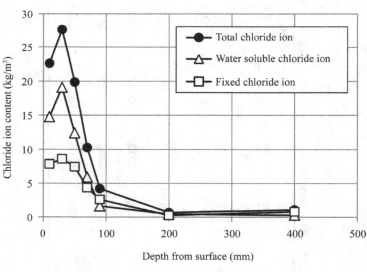

(b) point C core (fresh water)

Figure 4.3 Distribution of Cl⁻ content in mortar of point B (seawater) and point C (freshwater).

(a) Mother vessel

(b) Working platform

Figure 4.4 Mother vessel and working platform.

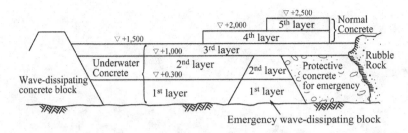

Figure 4.5 Cross section of concrete walls for footing protection [1].

The lifting cradle had not only a concrete pump and a seawater server for the proper casting of the concrete but also a drainage equipment incorporating a settlement tank and a pH control system to prevent marine pollution. When designing the concrete mix proportions, time-dependent changes of slump flow were investigated in detail, because the time from mixing to placement was over 2 hours and the air temperature was 30°C or more. Temperature stress analysis was also carried out with respect to thermal cracking. Inert aggregate and GGBFS powder were used to inhibit the alkali–silica reaction. Although wave forces act on the exposed concrete, horizontal joints between the concrete layers were unavoidable as shown in Figure 4.5. In addition, thermal cracking was a concern because of the large area of disc-shaped concrete. Therefore, the following countermeasures were taken:

1) Reinforcing bars were placed across horizontal joints to provide shear resistance by dowel action. Also, epoxy-coated bars were used in the seawater concrete.
2) Joint plates were placed into the disc-shaped concrete to control cracking.

BIBLIOGRAPHY

1. The Ministry of Construction (at present, Ministry of Land, Infrastructure, and Transport, and Tourism). Kanto Regional Development Bureau, Keihin Work Office. Record of Okinotorishima Island disaster-relief work (In Japanese).

4.3.2 Blocks Manufactured with Seawater Concrete for Soma Port

This section describes an example of concrete wave-dissipation and footing protection blocks for Soma Port, Fukushima Prefecture, Japan. The blocks were fabricated using seawater and as much uncrushed concrete debris as possible, aimed at the reutilization of concrete debris from earthquake-damaged concrete structures (Figure 4.6) [1, 2]. In order to produce concrete using uncrushed concrete debris in large sizes, both the pre-packed and post-packed concrete methods were used. Additionally, by using seawater as the mixing water, it was expected to reduce construction time and improve durability through the early and long-term strength development of the concrete [3].

(I) Fabricated Structure and Construction Method

Wave-dissipating blocks and footing protection blocks for ports and harbors are both of unreinforced concrete with a design strength of 18 N/mm². The post-packed concrete method was adopted for the

Figure 4.6 Concrete debris from the 2011 Great East Japan Earthquake disaster.

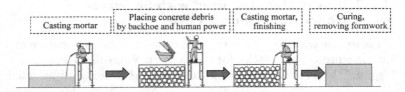

Figure 4.7 Post-packed concrete method.

footing protection blocks (Figure 4.7) and the pre-packed concrete method for the wave-dissipating blocks. For comparison, the same blocks were also produced by mixing concrete using crusher-run aggregate (maximum size 40 mm) made from concrete debris as coarse aggregate (termed 'recycled concrete'). A site plant was built to mix the mortar and concrete.

(2) Mix Proportions and Quality Control

For the designed performance of the mortar, the flow time was set at 90 seconds or less (JSCE F 521 'Test method for flowability of grout mortar for prepacked concrete'). The air content of the mortar was set at 8–12% (JIS A 1128 'Method of test for air content of fresh concrete by pressure method'), and the air content of the concrete was set at $5.5 \pm 1.5\%$, considering that the concrete blocks would be subject to freezing and thawing. And, the expansion ratio of mortar was set at 2–5% (JSCE F 522 'Test method for bleeding ratio and expansion ratio of grout mortar for prepacked concrete (Polyethylene bag method)').

Table 4.3 describes the materials used for the mortar and concrete. The origin of the concrete debris was concrete caissons damaged by the tsunami disaster and then broken up using concrete breakers, etc. The properties of the concrete debris are as follows: particle diameter was 300–500 mm, density was 2.37 g/cm³, absorption was 7.18%, and compressive strength was 37.2 N/mm².

Portland blast furnace slag (BFS) cement type B was used. Seawater or freshwater was used as the mixing water for the mortar, and freshwater was used as the mixing water for the concrete. In order to suppress shrinkage cracks, an expansive additive was used, while to ensure the integrity of the mortar and concrete debris, a foaming agent (aluminum powder) was used. Table 4.4 presents the mix proportions of the mortar and concrete. The W/B

Table 4.3 Materials of mortar and concrete

Material	Description	Code	Summary specification
Water	Freshwater	W	Tap water
	Seawater		Chloride ions 1.88%
Binder (B)	Portland blast furnace slag cement type B	C	Density 3.04 g/cm^3
	Expansive additive	Ex	Major component CaO, Density 3.16 g/cm^3
Coarse aggregate	Recycled crusher-run from concrete debris	G	Density 2.20 g/cm^3, Absorption 12.4%, Particle diameter 40 mm or less
Fine aggregate	Crushed sand	S	Density 2.66 g/cm^3, Particle diameter 5 mm or less
Foaming agent	Aluminum powder	Al	Reaction retarding type

of the mortar for the wave-dissipating blocks was 40.0%, and that for the foot-protection blocks was 45.0%. The amount of aluminum powder needed to satisfy the target expansion ratio was 40 g/m^3, replacing cement in the mortar.

Compressive strength was evaluated using two types of specimens. One type was cast in a steel mold (φ150 × 300 mm height) by the pre-packed and post-packed concrete methods using concrete debris of particle size about 40 mm (hereinafter called the φ150 mm specimens). The other type comprised core samples (φ150 × 300 mm length) from concrete blocks (cube side = 800 mm) made by the pre-packed and post-packed concrete methods using concrete debris of particle size of 300–500 mm (hereinafter called the core specimens; Figure 4.8).

(3) Concrete Properties

Figure 4.9(a) shows how the blocks were cast and Figure 4.9(b) shows the finished blocks after the formwork was removed. Figure 4.10(a) and 4.10(b) show the results of compressive strength tests. The compressive strength of the freshwater concrete exceeded the design strength (18 N/mm^2) at 28 days. On the other hand, the seawater concrete exceeded the design strength at 7 days. At 28 days, the compressive strength of the

Table 4.4 Mix proportions and properties of mortar and concrete

| Type | Mixing water | W/B (%) | S/B | s/a (%) | Unit weight (kg/m³) | | | | | | Flow time (second) | Slump (cm) | Air content (%) |
| | | | | | W*1 | B | | S | G | Al | | | |
						C	Ex						
Mortar for pre-packed concrete	Sea water	40.0	1.7	–	263	618	40	1119	–	0.04	58.3	–	10.0
	Freshwater	40.0	1.7	–	263	618	40	1119	–	0.04	49.2	–	10.7
Mortar for post-packed concrete	Sea water	45.0	1.7	–	286	595	40	1080	–	0.04	33.0	–	9.0
	Freshwater	45.0	1.7	–	286	595	40	1080	–	0.04	31.4	–	9.7
Recycled concrete	Freshwater	54.7	–	46.4*²	175	320	–	269	1303	–	–	11.0	5.6

*1: Assuming the density of seawater as 1.00 g/cm³ (actually 1.03 g/cm³)

*2: Counting 35% of recycled crusher-run (particle diameter 5 mm or less) fine aggregate

(a) Casting block (pre-packed concrete) (b) Drilled core

Figure 4.8 **Manufacturing of core specimens.**

(i) Wave-dissipating block (pre-packed concrete) (ii) Foot protection block (post-packed concrete)

(a) Casting of blocks for Soma Port

(i) Wave-dissipating block (pre-packed concrete) (ii) Foot protection block (post-packed concrete)

(b) Finished blocks for Soma Port

Figure 4.9 **Casting of blocks and finished blocks of Soma Port.**

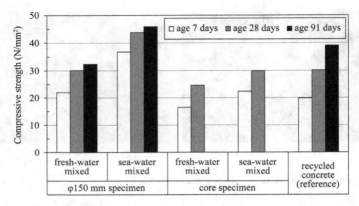

(a) Compressive strength of wave dissipating blocks (pre-packed concrete)

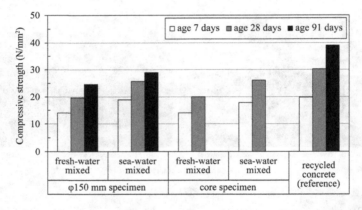

(b) Compressive strength of footing protection blocks (post-packed concrete)

Figure 4.10 Compressive strength of wave-dissipating blocks (pre-packed concrete) and footing protection blocks (post-packed concrete).

wave-dissipating blocks (W/B = 40.0%; seawater) was 30 N/mm²
or higher, and that of the foot-protection blocks (W/B = 45.0%;
seawater) was 25 N/mm² or higher. For both the wave-dissipating
blocks and the footing protection blocks packed with seawater
mortar, the compressive strengths were higher than those made
with freshwater mortar.

Figure 4.11 Batch mixer for seawater concrete.

4.3.3 Mixer Condition after Seawater Concrete Mixing

Figure 4.11 shows the concrete mixer used at the on-site plant over a three-month period to manufacture mortar using seawater. The batch mixer was kept rust-free by washing with freshwater once a day.

BIBLIOGRAPHY

1. Katano, K., Takeda, N., Hisada, M. and Otsuki, N. 2013. Properties of prepacked concrete using seawater and concrete debris generated by the earthquake disaster. The 68th JSCE Annual Meeting:17–18.
2. Yatabe, M., Hisada, M., Otsuki, N. and Kanbara, S. 2013. A test construction of prepacked concrete using seawater and concrete debris generated by the earthquake disaster. The 68th JSCE Annual Meeting:19–20.
3. Takeda, N., Ishizeki, Y., Aoki, S. and Otsuki, N. 2011. Development effect of strength and watertightness by using sea water as mixing water for concrete. The 66th JSCE Annual Meeting:581–582.

Figure 1.10 Interior view of the mixer boxes plate.

1.3.3 Mixer Condition after Seawater Contamination

Figure 1.10 shows the condition only minutes after the mixer plant was run a time through, as had it would enough mortar after using seawater. The batch were washed just five minutes along with fresh water enough during.

BIBLIOGRAPHY

1. Kumar, S., Prasad, N., Joseph, M. and Vivek, N. (2018). Properties of compacted concrete using seawater and cement materials accepted by the third publication. Proceedings of 36th Annual Meeting of the Transportation Mumbai, Otto, L. S., and Joseph, S. 2019. A cost-efficient air-prestressed concrete using seawater and concrete structure or fine aggregate in cement. Proceedings of 68th ISC Annual Meeting, 2019.

2. Takehira, M., Nakagi, C., Sekhar, S. and Waik, P. 2019. Development of precast concrete structure and medium. Evaluation of water mixing using marine concrete. 22th annual Annual Meeting of C.S. 2018.

Chapter 5

European Experience with Seawater Concrete

5.1 INTRODUCTION

The use of seawater for concrete production is not common in Europe now, even for the production of plain concrete exposed to a marine environment. However, in recent decades, certain international and national seawater concrete research projects have been initiated in Europe, including the Roman Maritime Concrete Study (ROMACONS), Sustainable Concrete using Seawater, Salt-contaminated Aggregates, and Non-corrosive Reinforcement (SEACON), and Eco-efficient By-products Suitable for the Market through the Integration of Recycling in Port Environments (SEA MIRENP). Both ROMACONS and SEACON were jointly carried out by European and non-European researchers. In both projects, actual-scale structures were analyzed. UPC, the researcher from Universitat Politeècnica de Catalunya (UPC.BarcelonaTECH), developed the SEA MIRENP project and carried out actual-scale tests at Barcelona Port.

This chapter outlines these three projects and describes the actual-scale application of concrete produced using seawater for the marine environment as implemented in Europe as part of the projects.

5.2 ROMAN SEAWATER CONCRETE

Roman maritime concrete structures have remained cohesive and intact for 2,000 years. The ROMACONS project, which ran from 2001 to 2009, conducted fieldwork along the Mediterranean Sea coast (Italy, Greece-Crete, Turkey, Israel, and Egypt) [1–3],

DOI: 10.1201/9781003194163-5

collecting cores from maritime structures built using the Roman hydraulic mortar or concrete. Figure 5.1(a) illustrates the locations where cores were collected from these marine structures.

A hydraulic-powered Cordiam Hydro 25 drill motor provided sufficient power to drill through at least 6 m of Roman concrete. The photos in Figure 5.1(b) show coring work in progress and one of the extracted cores.

(a) Locations of maritime structures constructed of Roman hydraulic mortar or concrete from which cores were taken [3]

(b) Extracting cores from a submerged block and exposed mole [3]

Figure 5.1 Locations of maritime structures constructed of Roman hydraulic mortar or concrete from which cores were taken and extracting cores from a submerged block and exposed mole [3].

The cores obtained from this fieldwork confirmed that the Roman engineers considered pozzolana (from Latin *pulvis*, lit. 'powdered sand') from Campi Flegrei (the Bay of Naples) to be the optimal ingredient for hydraulic concrete in maritime structures both in Italy and elsewhere in the Mediterranean [2, 3]. Volcanic ash from Campi Flegrei and other Vesuvius deposits has a composition unusually high in alkali content and low in silica (up to 12 weight% $Na_2O + K_2O$ and rich in aluminosilicates) [3]. Roman hydraulic concrete consisted of a mortar made from lime, pozzolana, and water with the addition of various types of rubble aggregate.

Analytical investigations of the ROMACONS samples indicate that seawater was an integral component of the concrete mix design [4–6]. It is conceivable that builders could have incorporated pumiceous ash pozzolana and matured pebble lime (or quicklime) in a mortar trough and then lowered this as a dry mix or as seawater-moistened clumps into forms to produce a mortar that fully hydrated *in situ*. The broken volcanic tuff or limestone *caementa* would then have been tossed in and the mixture compacted. The resultant mixture was a hard and durable concrete that could solidify underwater.

The researchers who worked on ROMACONS also constructed a reproduction of a Roman maritime structure (a *pila*) in Brindisi, Italy [7]. The maritime mortar mixture was produced employing pozzolana (*pulvis*), lime, and a small amount of water [3, 7].

A theoretical understanding of Roman hydraulic concrete was taken from the only surviving ancient handbook of architecture, *De Architectura* by Vitruvius, which was probably published between 30 and 22 BC [3]. Neither Vitruvius nor any other source goes into great detail about procedures for measuring and mixing the ingredients of the hydraulic mortar, the use of salt or freshwater, or the method used to place mortar and aggregate in inundated forms. Since Vitruvius is careful to specify the source and quality of the pozzolana and lime that constitute the mortar, he would almost certainly have indicated the use of freshwater for mixing concrete destined for maritime structures if the nature of the water had been an issue. Besides, the provision of large quantities of freshwater at a harbor construction site usually would have involved serious logistical problems. Seawater and natural pumiceous ash pozzolana were also used, for example, in concrete produced for both the Suez and Corinth canals [3]. In addition, the analytical investigations of the ROMACONS samples indicate that seawater was an integral component of the concrete mix design [5, 6, 8].

In the reproduction carried out as part of the ROMACONS project [3, 7], the hydraulic pozzolanic mortar produced was composed mainly of hydrated lime and pumiceous volcanic ash (*pulvis*). The proportions (by volume) of 2 or 2.5 parts of pozzolana to 1 part lime and a small amount of water were employed, producing a stiff mix with little or no slump. It is imagined that in Roman times baskets would have been filled at a mixing trough, then carried to the edge of the form by one or two individuals who lowered them to the selected spot with the two rope handles before emptying the contents by means of a pull rope. The reproduction required 356 basket loads of mortar, each followed by the addition of aggregate, to fill the formwork [3, 7]. This reproduction of Roman concrete was tamped after each day's placing using a long-handled rake to spread the mortar into the corners and compact it across the form. Compaction of the mortar tended to release numerous rounded pumice *lapilli*, inclusions in the pozzolana that did not bond with the mortar in the course of mixing. Inclusions of this type were frequently seen in the cores taken from Roman maritime structures.

The visual appearance of cores extracted from the reproduction block was remarkably similar to those collected from the ancient Roman structures. Voids appear with slightly greater frequency in the reconstructed concrete than in the ROMACONS core samples, suggesting that the ancient concrete may have been tamped more frequently or more thoroughly, or placed with greater skill [3].

Recent nanoscale investigations of the cementitious components of the ancient seawater concrete [5, 6, 8, 9] report that the extraordinary longevity of the concrete seems to result from the long-term durability of poorly crystalline calcium-aluminium-silicate-hydrate (C-A-S-H) binder in the cementing matrix of the mortar, the sequestration of chloride and sulfate ions in discrete crystalline microstructures (hydrocalumite and ettringite) and pervasive crystallization of zeolite and Al-tobermorite mineral cements in pumice clasts, dissolved feldspar crystal fragments, and relict voids of the cementing matrix.

According to several researchers [3, 5, 6], the dissolution of the solid aluminosilicate occurs through alkaline hydrolysis, which releases aluminate and silicate species that are incorporated into an aqueous phase. Concentration of these species in solution produces a gel that continues to reorganize and transform and that grows in connectivity to form three-dimensional polymerized networks that are responsible for hardening the concrete. This is, essentially, the hardening process that Vitruvius describes empirically for the consolidation of the seawater concretes in *De Architectura* and which is recorded by the C-A-S-H binding phase of the cementitious matrix. Alkali cations, mainly sodium but also potassium, greatly assist the process of dissolution; OH⁻ ions act as a reaction catalyst for the formation of aluminous cementitious gel, while alkali cations act as structure-forming elements to balance Al^{3+} substitution for Si^{4+}. Aluminium substitution for silica may be an important factor in the chemical durability of the ancient C-A-S-H binder and Al-tobermorite. The charge balance introduced by Al^{3+} substitution for Si^{4+} encourages the binding of alkali cations and seems to contribute to equilibrium in the seawater concrete environment. These reactions do not occur in conventional concretes, so these aluminous, cementitious hydrates hold great potential as binders for environmentally sustainable concretes and concrete encapsulations of hazardous and nuclear wastes.

BIBLIOGRAPHY

1. Hohfelder, R.L., Brandon, C. and Oleson, J.P.. 2007. Constructing the harbour of Caesarea Palaestina, Israel. New evidence from the ROMACONS field campaign of October 2005. *International Journal of Nautical Archaeology*, Vol.36, No.2:409–415.
2. Brandon, C.J., Hohlfelder, R.L. and Oleson, J.P.. 2008. The concrete construction of the Roman harbours of Baiae and Portus Iulius, Italy. The ROMACONS 2006 field season. *International Journal of Nautical Archaeology*, Vol.37, No.2:374–379.
3. Brandon, C.J., Hohlfelder, R.L., Jackson, M.D. and Oleson, J.P.. 2014. *Building for Eternity*. Oxford: Oxbow Books.
4. Monteiro, P. and Jackson, M.. 2013. To improve today's concrete, do as the Romans did. [Online] Available at: https://newscenter.lbl.gov/2013/06/04/roman-concrete/
5. Jackson, M.D. et al. 2014. Cement microstructures and durability in ancient Roman seawater concretes. In RILEM Bookseries, Vol.7, No. June.
6. Jackson, M.D. et al. 2017. Phillipsite and Al-tobermorite mineral cements produced through low-temperature water-rock reactions in Roman marine concrete. *American Mineralogist*, Vol.102, No.7:1435–1450.

7. Oleson, J.P., Bottalico, L., Brandon, C., Cucitore, R., Gotti, E. and Hohlfelder, R.L.. 2006. Reproducing a Roman Maritime structure with Vitruvian pozzolanic concrete. *Journal of Roman Archaeology*, Vol.19, No.1:29–52.
8. Monteiro, P. and Jackson, M.. 2013. To improve today's concrete, do as the Romans did.
9. Jackson, M.D. et al. 2012. Unlocking the secrets of Al-tobermorite in Roman seawater concrete. *American Mineralogist*, Vol.98, No.10:1669–1687.

5.3 SUSTAINABLE CONCRETE USING SEAWATER, SALT-CONTAMINATED AGGREGATE, AND NON-CORRODING REINFORCEMENT

The SEACON project started in 2015 to promote the best practice in producing concrete and reinforced concrete structures using alternative materials [1, 2]. The aim was to reduce the use of critical resources by replacing them with chloride-contaminated alternatives coupled with non-corroding reinforcement.

An actual-size field prototype, a concrete culvert, was constructed in Italy [3, 4]. The culvert, parallel to a motorway, was designed for the collection of drainage water. De-icing salts are used on the highway during the winter. Further, the culvert being unsheltered, it is exposed to wetting and drying cycles. Figure 5.2 shows the culvert scheme with its six segments (A–F) using different concretes and reinforcements, and a cross-section.

Table 5.1 shows the mix proportions of the concretes used and their fresh-state properties. The fresh-state properties of the three types of concrete were similar. The type of concrete and reinforcement employed in each of the culvert's six sections are described in Table 5.2. Figure 5.3 shows the data acquisition system used in all the segments of the culvert.

The hardened-state properties are described in Table 5.1. Although the RAP concrete achieved the lowest density, the densities of the three concretes can be considered similar. The concrete produced with seawater (SEACON concrete) achieved the highest compressive strength, followed by the reference concrete. The RAP concrete reached the lowest compressive strength value.

According to the researchers [5], the potential of the reinforcement in all the segments was very negative, in particular for the

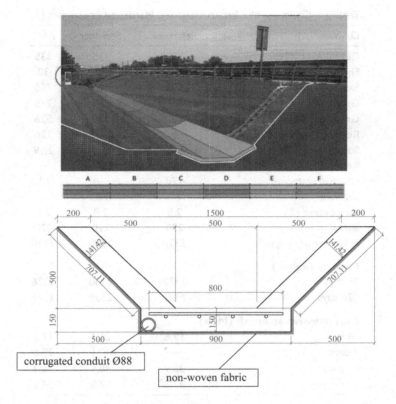

Figure 5.2 Culvert scheme and cross section [3].

carbon steel reinforcement. After 8 months of exposure, the potential of the carbon steel had increased to −50 mV versus the embedded silver/silver chloride electrode (SSC), −100 mV versus SSC and −200 mV versus SSC for the reference, RAP and seawater concrete, respectively. The stainless-steel reinforcement showed a slight increase, up to −24 mV versus SSC. These preliminary results highlight the different behavior of, on the one hand, carbon steel in the reference concrete, RAP concrete, and stainless steel in seawater concrete, and on the other hand, carbon steel in seawater concrete. The latter is characterized by lower potential values compared to the other concretes. This behavior seemed to suggest that carbon steel in seawater concrete approaches a condition of depassivation. Further measurements are necessary to confirm this hypothesis.

Table 5.1 Concrete Mix Proportions and Hardened Properties [3]

Concrete	Reference	Seawater	RAP*
CEM II/A-LL 42.5R	335	335	335
Fly ash	30	30	30
Sand 0–5 mm	800	800	766
Gravel 5–7 mm	365	365	246
Gravel 8–15 mm	630	630	526
Recycled aggregate (RAP)	–	–	226
Superplasticizer	2.19	2.19	2.19
Retarding agent	–	0.76	–
Effective water	178	–	173.3
Seawater	–	176.7	–
Air content (%)	2.5	2.8	2.8
Slump (mm)	230	220	220
Fresh density (kg/m³)	2320	2340	2330
Density (kg/m³)			
28 days	2372	2380	2336
120 days	2381	2389	2372
Compressive strength (MPa)			
24 h	19.6	21.1	17.2
7 days	36.7	40.1	28.6
28 days	38.9	48.6	36.2
120 days	47.0	57.5	34.5

Figure 5.4 shows the concrete resistivity measured with the embedded resistance probe at the depth of the reinforcement as a function of time [5]. The resistivity values of the six segments were similar at an early age. However, after eight months, the concretes produced with seawater had achieved a higher resistivity value. Although these results were opposite to those obtained in a laboratory test, in the data obtained so far, it is clear that seawater concrete achieves a higher resistivity. It is probable that the higher compressive strength reached by concrete produced with seawater could influence this property.

The chloride penetration value [4] was negligible during the first year of exposure, and the chloride content at the depth of the rebar was the same as initially (except for the RAP concrete, where a slight increase was observed).

Table 5.2 Types of Reinforcement, Concrete, and Measurement Systems Present in Culvert Segments [3]

Segment	A	B	C	D	E	F
Reinforcement	Carbon Steel	Carbon steel	Stainless steel 304	Stainless steel 23-04	GFRP	Carbon steel
Concrete	Reference	Seawater	Seawater	Seawater	Seawater	RAP
Rebar	⊻	⊻	⊻	⊻	–	⊻
SSC-Ref	⊻	⊻	⊻	⊻	–	⊻
Ri-Ref	⊻	⊻	⊻	⊻	⊻	⊻
Res-Probe	⊻	⊻	⊻	⊻	⊻	⊻
Ti-mesh	⊻	⊻	–	–	–	⊻
Multi-probe	⊻	⊻	–	–	–	⊻

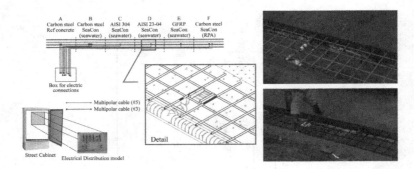

Figure 5.3 Data acquisition system and positioning of probes and electrodes [3].

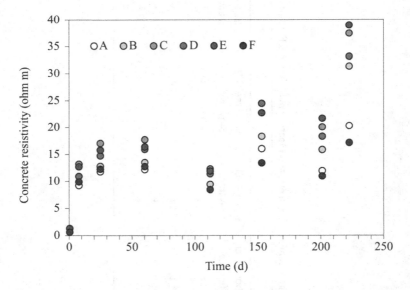

Figure 5.4 Concrete resistivity at the depth of the reinforcement as a function of time [5].

The conclusion from the preliminary results after eight months was that carbon steel seawater concrete was characterized by more negative potential values. Overall, the results will be better interpreted after monitoring for a longer duration and in combination with other measurements. (The work is ongoing.)

BIBLIOGRAPHY

1. Bertola, F., Gastaldi, D., Canonico, F. and Nanni, A.. 2017. SEACON Project: sustainable concrete using seawater, salt-contaminated aggregates, and non-corrosive reinforcement. In XIV DBMC – 14th International Conference on Durability of Building Materials and Components.
2. University of Miami (Coord. Nanni, A. U.M.) and Politecnico de Milano (POLIMI) and Buzzi Unicem. 2018. SEACON Report. [Online], Available at: http://seacon.um-sml.com/uploads/1/6/7/2/16727926/d7.3_seacon_final_report.pdf
3. Buzzi Unicem and Nanni, A. (Project Coordinator). 2017. Field demo of culvert in Italy. Deliverable D4.2.
4. Bertona, F., Canonico, F., Redaelli, E., Carsana, M., Gastaldi, M., Lollini, F., Torabian Isfahanu, F. and Nanni, A.. 2020. On-Site demonstration project of reinforced concrete with seawater. *Lecturer Notes in Civil Engineering*, Vol.42:474–486.
5. Buzzi Unicem and Nanni, A. (Project Coordinator). 2017. Long-term on-site monitoring of field demos. Vol.113. Available at: http://seacon.um-sml.com/uploads/1/6/7/2/16727926/d4.4_long-term_onsite_monitoring.pdf

5.4 SUSTAINABLE CONCRETE DYKE BLOCK PRODUCTION IN BARCELONA [1]

During July 2014, several dyke blocks were produced in Barcelona employing seawater (SW) and secondary aggregates in order to validate the viability and sustainability of the product under the SEA MIRENP project. Coarse mixed recycled aggregates (RA) and coarse steel slag aggregates (S) were employed as secondary aggregates. All the concretes were produced employing natural sand.

The results obtained for these dyke block concretes using seawater and secondary aggregates were compared to results obtained with conventional concrete (using natural aggregates and freshwater, W). In total, seven cubic blocks of concrete were fabricated. Their mix proportions are described in Table 5.3. Figure 5.5 shows the production process and images of the blocks. The sides of each block were 2.8 m long.

In order to characterize the concrete of these dyke blocks, specimens were produced with the same concretes, and these were tested up to one year of age. Cores were extracted from the dyke

Table 5.3 Concrete Dosing, in Units of kg per m³ of Concrete [1]

Dosing according to PROMSA	CEM I 42.5/N SR	Sand 0/4 (mm)	Gravel 4/10 (mm)	Gravel 10/20 (mm)	Recycled aggregate 5/20 (mm)	Steel slag 5/10 (mm)	Steel slag 10/20 (mm)	Effective or water	Total water	Effective W/C	Admixture (%)	Slump (cm)
Block 0 CC-W	300	976	210	765	–	–	–	134	150	0.446	1	10
Block 1 CC-SW	300	976	210	765	–	–	–	153	169	0.509	1.11	7
Block 2 CRS-W	300	976	–	–	385	143	506	145	177	0.484	1.16	10
Block 3 CRS-SW	300	976	–	–	385	143	506	157	189	0.524	1.21	7
Block 4 CRA-50-W	300	976	105	383	385	–	–	143	175	0.477	1.16	10
Block 5 CRA-50-SW	300	976	105	383	385	–	–	149	181	0.497	1.26	10
Block 6 CRA-100-SW	335	826	–	–	889	–	–	151	201	0.451	1.26	10

CC: conventional concrete; CRS: concrete produced with 50% RA and 50% S; CRA-50: concrete produced with 50% RA; CRA-100: concrete produced with 100% RA.

Figure 5.5 Concrete dyke blocks and extracted cores [1].

blocks after one year. The results for the cores and specimens are compared.

Table 5.4 shows the physical properties of the concrete specimens at 28 days. The specimens made with seawater had a higher density than those made with freshwater. Those specimens made with steel slag concrete had the highest density due to the slag aggregate, while the others (produced with RA) had a lower density than conventional concrete.

Table 5.4 shows the mechanical properties after 7, 28, 90, 180 days, and 1 year of curing. Results show that the compressive strength of concrete produced with secondary aggregate and seawater is lower than conventional concrete in all cases. However, the obtained results are higher values than the minimum required by the port authority for dyke blocks, except for Block 6. The specimens made with seawater had similar or slightly lower compressive strength than those produced with freshwater. Table 5.3 shows that the seawater concrete was produced with a higher effective W/C ratio. More water was needed because seawater causes a faster setting time. The splitting tensile strength of conventional

Table 5.4 Physical Properties, Mechanical and Durability Properties of Concrete Specimens and Properties of Extracted Cores [1]

Test Period	CC		CRS		CRA-50		CRA-100	
	-W Block 0	-SW Block 1	-W Block 2	-SW Block 3	-W Block 4	-SW Block 5	-SW Block 6	Port*
PHYSICAL PROPERTIES (28 days)								
Dry density (kg/dm³)	2.34	2.36	2.40	2.43	2.24	2.25	2.15	2.2
Water absorption (%)	4.04	3.30	4.96	4.58	5.07	4.94	5.67	–
Porosity (%)	9.47	7.77	11.89	11.14	11.36	11.12	12.19	–
COMPRESSIVE STRENGTH (MPa) (cubic specimens)								
28 days	46.97	45.56	36.79	35.76	39.07	36.05	29.21	30
1 year	58.9	60.2	50.3	43	42.5	42.5	36.0	
SPLITTING TENSILE STRENGTH (MPa)								
28 days	3.92	3.29	2.83	3.09	2.47	2.50	2.39	
PHYSICAL PROPERTIES of CORES								
Density (kg/dm³)	2.33	2.33		2.34	2.22	2.20	2.08	
Absorption (%)	4.83	5.07		6.75	6.66	6.66	9.13	
COMPRESSIVE STRENGTH (MPa) of CORES								
	53.8	51.3		36.4	38.6	37.5	29.9	30
WATER PENETRATION MAXIMUM VALUE (mm) of CORES								
Maximum	25	35		27	32	25	36	50
Average	15	23		19	22	17	28	–

*Port requirements (minimum density of 2.2–2.3 kg/dm³).
*Port requirements (minimum compressive strength of 30 MPa). Spanish standard for structural concrete (EHE) maximum requirement for durable concrete.

concrete produced with freshwater was higher than that produced with seawater, probably due to the higher effective W/C ratio used in seawater concrete production. The concrete produced with RA and S aggregates employing seawater achieved a higher splitting tensile strength than those produced with freshwater.

The durability properties of all the concretes were tested at 28 days and one year: capillary water absorption coefficient, electrical resistivity, and water penetration. It was observed that concrete produced with seawater achieved a lower capillary absorption coefficient and water penetration value than those produced with freshwater. On the contrary, the electrical resistivity of concrete produced with seawater was lower than that of those produced with freshwater.

After the blocks had been exposed to a marine environment for one year, cores were extracted from all blocks except Block 2, which was inaccessible. The hardened state properties of the blocks are shown in Table 5.4. The seawater concretes achieved similar physical properties to the concretes produced with freshwater. In contrast, the compressive strength of seawater concretes was a little lower than that of freshwater concretes but, as mentioned above, this was probably due to the higher effective W/C in concretes produced with seawater. All the concretes produced with seawater exhibited lower water penetration than the corresponding concrete (according to the type of aggregates used) made with freshwater.

The authors concluded [1] that the use of seawater in concrete production reduced the setting time compared to that of the freshwater concrete. In addition, density values were higher while values of porosity and water penetration were lower. The mechanical properties of the concretes were influenced more by the type of aggregate than the type of water employed.

BIBLIOGRAPHY

1. Etxeberria, M., Fernandez, J. M. and Limeira, J.. 2016. Secondary aggregates and seawater employment for sustainable concrete dyke blocks production: Case study. *Construction and Building Materials*, Vol.113:586–595.

Chapter 6

Summary and Future Scope

6.1 TECHNICAL OPINION OF AUTHORITIES ON CONCRETE

6.1.1 Professor (Dr.) Adam Neville [1]

Prof. Neville's views are explained in detail in the reference below, which compiles his opinions on the effect of seawater in concrete from the time of mixing. Presently, the common belief in western countries regarding the use of seawater as mixing water in concrete appears to be that it is forbidden in reinforced concrete, absolutely forbidden in precast concrete, and prohibited as far as is possible in unreinforced concrete. Prof. Neville questions this stereotypical thinking. In fact, in the past literature, there are papers with arguments for permitting the use of seawater, Thus, 'keeping ajar the door to the use of seawater.' There is no need to completely close the options for the possibility of seawater kneading. However, as it is suspected to cause severe deterioration depending on the environment of use of the concrete, it is necessary to exercise caution in its application. Prof. Neville takes a tough stand in the matter of using seawater for curing. In his opinion, seawater curing must be prohibited even if the structure is used in a seawater environment, because according to him, Cl^- ions enter the concrete with considerable swiftness during the strength development process. Prof. Neville's basic idea is that seawater should not generally be used for mixing concrete, but he makes two exceptions: (1) when the concrete is to remain in a completely dry state during its period of service and (2) when it is placed within seawater or is completely immersed in water. He has opined that only in these two cases may seawater be used for mixing. It is mentioned at the end of his report that this restriction does not apply to unreinforced concrete.

BIBLIOGRAPHY

1. Adam Neville. 2001. Seawater in the mixture. *Concrete International*, Vol.23, No.1, January, pp.48-51.

6.1.2 Professor (Dr.) Yuzo Akatsuka [1]

Prof. Akatsuka's views are explained in detail in the reference below. Here, we summarize his views on the use of seawater for mixing concrete. He states that opinions are divided among engineers on the use of seawater for mixing concrete. Many would permit the use of seawater for unreinforced concrete, whereas many would prohibit its use for reinforced concrete as would approve of it. His opinions can be summarized as follows. Regarding the effect on strength and setting time, it is reasonable to regard the differences to be not substantial enough to consider freshwater or seawater as alternatives for mixing concrete when designing admissible stress. As for its effect on durability, there are several examples of port, coastal, and lighthouse structures built using unreinforced concrete mixed with seawater that have been in service for several decades with practically no degradations despite being under the influence of seawater. As to the effect on rebar corrosion, the German standards (DIN) permit the use of seawater. The United States has rules that specifically prohibit the use of seawater (American Concrete Institute, ACI) and also permit its use (Portland Cement Association, PCA) under different circumstances. In the UK (British Standards, BS), seawater is prohibited for only those structures that are subjected to the action of seawater. Prof. Akatsuka's view is that although the use of seawater for mixing cannot be encouraged, there appears to be no reason for absolute prohibition either. When there are some overlapping defects in the design and construction, taking appropriate action with due caution is necessary, such as adequate measures regarding fracturing, rebar corrosion, cover thickness (making the mix particularly water-tight), strict construction control, and so on. In general, there are very few instances in which the use of seawater for mixing concrete is inevitable; except for the construction of breakwaters and lighthouses on remote islands and in offshore locations, and for seawall construction work in coastal areas with poor freshwater availability, it is generally rare for a difficulty to arise in obtaining fresh water for harbor work. In the exceptional case where seawater has to be used, it would be prudent to carefully control the construction work, considering the points discussed above, rather than spending vast amounts on the transport of freshwater.

BIBLIOGRAPHY

1. Yuzo Akatsuka. 1977[1969]. *Concrete Harbor Structures, Concrete Books*, No.14. Institute of Cement Association Publication, First Edition, June 20, 1969, Enlarged and Revised Edition, January 20, 1977.

6.2 RECENT RESEARCH ON SEAWATER IN THE CONCRETE MIX

The following list comprises recent doctoral and master studies on the subject completed at Kyushu University.

6.2.1 Dr. Adiwijaya [1–4]

Dr. Adiwijaya's research focused on the physical properties of concrete using seawater as mixing water. One very interesting and important result is on the alkali–silica reaction (ASR) reactivity of seawater concrete. Due to his research results, concrete with ordinary Portland cement (OPC) expands due to the ASR reaction when seawater is used as mixing water. However, concrete, including Blast Furnace Slag (BFS) or Fly Ash (FA), does not expand even if seawater is used as mixing water. Thus, as a preventive measure against ASR, mineral admixtures play an important and effective role.

BIBLIOGRAPHY

1. Adiwijaya. 2015. *A fundamental study on seawater-mixed concrete related to strength, Carbonation and Alkali Silica reaction*, PhD diss., Kyushu University. https://catalog.lib.kyushu-u.ac.jp/opac_download_md/1543974/eng2500.pdf

2. Adiwijaya, Hidenori Hamada, Yasutaka Sagawa and Daisuke Yamamoto. 2015. Expansion characteristics of seawater mixed concrete due to Alkali-Silica reaction. The 40th Conference on Our World in Concrete and Structures (OWICS):311–320.

3. Adiwijaya, Hidenori Hamada, Yasutaka Sagawa and Daisuke Yamamoto. 2014. Effects of mix proportion and curing condition on Carbonation of Seawater-Mixed concrete. The Conference for Civil Engineering Jointly held with The 7th ASEAN Civil Engineering Conference (ACEC):64–69.

4. Adiwijaya, Hidenori Hamada, Yasutaka Sagawa and Daisuke Yamamoto. 2014. Effects of mineral admixtures on strength characteristics of concrete mixed with seawater. The 6th International Conference of Asian Concrete Federation (ACF):925–930.

6.2.2 Dr. Amry Dasar [5–8]

Dr. Dasar's research includes an investigation of the electrochemical characteristics of steel reinforcing bars in seawater concrete up to five years after concrete production. Electrode potential in the very initial stage is rather low, indicating corrosion. However, it recovers gradually over time, and reaches a very stable condition beyond the age of six months. He also tested the corrosion resistance of epoxy-coated and stainless steel reinforcing bars in seawater concrete.

BIBLIOGRAPHY

5. Amry Dasar. 2017. *A study on deterioration and corrosion behavior of RC and PC members with initial defects under environmental action*, PhD diss, Kyushu University. https://catalog.lib.kyushu-u.ac.jp/opac_download_md/1866287/eng2684.pdf

6. Amry Dasar, Dahlia Patah, Hidenori Hamada, Yasutaka Sagawa and Daisuke Yamamoto. 2020. Applicability of seawater as a mixing and curing agent in 4-year-old concrete. *Journal of Construction and Building Materials*, Vol. 259:119692. https://doi.org/10.1016/j.conbuildmat.2020.119692

7. Amry Dasar, Hidenori Hamada, Yasutaka Sagawa and Daisuke Yamamoto. 2016. Recovery in mix potential and polarization resistance of steel bar in cement hardened matrix during early age of six months [Sea-water mixed mortar and cracked concrete]. *Proceedings of Japan Concrete Institute*, Vol.38, No.1:1203–1208. http://data.jci-net.or.jp/data_html/38/038-01-1196.html

8. Amry Dasar, Hidenori Hamada, Yasutaka Sagawa and Rita Irmawaty. 2013. Corrosion evaluation of reinforcing bar in sea water mixed mortar by Electrochemical method. *Proceedings of Japan Concrete Institute*, Vol.35, No.1:889–894. http://data.jci-net.or.jp/data_pdf/35/035-01-1144.pdf

6.2.3 Dr. Dahlia Patah [9,10]

Dr. Patah's research includes a discussion on the chloride threshold for steel corrosion in concrete with several different kinds of mineral admixtures: BFS, FA, silica fume, and Metakaolin. In her experiments, the threshold chloride content for steel corrosion is two to three times higher than that of concrete made with OPC only. These chloride contents are above the estimated level of chloride ions in seawater concrete. This means that steel reinforcement in seawater concrete with an adequate level of mineral admixtures should be stable without corrosion. This reflects the results of the literature review by Dr. Takahiro Nishida as reported in Chapter 1, in which results with seawater concrete specimens containing mineral admixtures tend to be positive, rather than negative.

Dr. Patah also carried out observations of old seawater concrete specimens with over 30 years' exposure to marine conditions. From her observations, it is shown that steel corrosion in seawater concrete is similar to or rather less than in freshwater concrete.

BIBLIOGRAPHY

9. Dahlia Patah. 2019. *A study on corrosion evaluation of steel reinforcement in concrete during initiation and propagation stage due to chloride attack*, PhD diss, Kyushu University. https://catalog.lib.kyushu-u.ac.jp/opac_download_md/2534439/eng2908.pdf
10. Patah, D., Hamada, H., Sagawa, Y. and Yamamoto, D. 2019. The effect of seawater mixing on corrosion of steel bar in 36-years old RC beams under marine tidal environment. *Proceedings of Japan Concrete Institute*, Vol.41, No.1:791–796.

6.2.4 Ms. Sabrina Harahap [11,12]

Ms. Harahap's research was briefly described in Chapter 3 of this book. Her idea is to protect steel in seawater concrete by coating the steel surface with a mortar mixed with a corrosion inhibitor. She began her experimental work on this idea and was able to continue testing for up to one year. Her experimental work is being followed up by another doctoral student (Ms. Volana Andriamisaharimanana), so tests are ongoing. Although limited to just a few years of results, so far, the pre-coating method shows good effectiveness at preventing corrosion of steel in seawater concrete.

Ms. Harahap also began research on the application of cathodic protection to steel in seawater concrete. This experiment is also being followed up by another student (Mr. K. Yoshida and Mr. M. Kai). In the first few years, cathodic protection has good protection effects against corrosion of steel in seawater concrete.

BIBLIOGRAPHY

11. Sabrina Harahap. *A study on corrosion behavior of steel reinforcement in seawater-mixed concrete and application of corrosion prevention – corrosion inhibitor and cathodic prevention*, Master Thesis, Kyushu University, Sept. 2018.
12. Sabrina Harahap, Hidenori Hamada, Yasutaka Sagawa and Daisuke Yamamoto. 2019. The effect of Calcium Nitrite coating as corrosion inhibitor in seawater-mixed mortar. The Proceedings of The 3rd ACF Symposium, Assessment and Intervention of Existing Structures.

6.3 CURRENT STATUS AND FUTURE SCOPE

Studies reported in the literature show that using seawater in concrete, both as mixing water and curing water, is feasible as long as certain practices are followed. In some cases, it can even be advantageous. For unreinforced concrete, its use is considered to be almost completely without problems, while for reinforced concrete it is suitable in almost all cases if adequate measures are devised to meet performance requirements. The approach to the use of seawater should be to focus on the ways to make such use possible, rather than on reasons why it is not possible. That is, to borrow a phrase, the attitude should be 'Where there's a will, there's a way.' Of course, the combination of poor concrete design (such as the large water-to-cement ratio) with a lack of special measures will result in significant chloride damage and is clearly out of the question; our fundamental knowledge of concrete engineering must be applied. The following is a summary of the main considerations for the use of seawater concrete.

Geographical constraints have, in the past, led to concrete structures being constructed using seawater and unwashed sea sand because it was often difficult to obtain freshwater. Many of these structures remain in service in a healthy state, giving us many hints about how seawater can be used. On the other hand, there are also structures and buildings with severe chloride damage resulting from the use of unwashed sea sand, and these structures remind us of the necessity for technically correct use. Studies of these existing structures tell us that it is possible to use seawater to build concrete structures with good performance if proper techniques are used. However, the present state of knowledge is barely sufficient, so the development of a technical methodology for the use of seawater is called for.

Looking at the physical properties of concrete using seawater as the mixing and curing water, strength development has been reported to be promoted during the initial aging and then to stagnate during the later aging in many cases. The reaction rate of each constituent mineral over time and the phase constitution of hydration products over time have been investigated to gain an understanding of the influence of seawater, resulting in new knowledge on the mechanism and kinetics of cement hydration. Such knowledge is essential to the proper use of seawater as mixing water and curing water in concrete. Mastery of these mechanisms will allow us to more confidently produce seawater concrete in the future.

The potential disadvantages of using seawater as mixing water have been confirmed: reinforcement corrosion, frost damage, and degradation of durability. However, by considering the type of cement, mix proportion and application environment, reinforcement corrosion can be suppressed. It has also been confirmed that frost damage can be suppressed by ensuring a proper amount of air, just as with freshwater concrete. Further, as regards the alkali–silica reaction, measures to control this should be taken, given that an alkaline component is introduced from the initial stage in the form of seawater. The durability of concrete using seawater as mixing water does not degrade immediately, and the required durability can be ensured by judiciously selecting the quality of the concrete in view of the environmental conditions.

Examples of the use of seawater to make concrete have long been reported, but in the vast majority of cases these uses were driven by the lack of freshwater availability. A more proactive use of seawater requires a proper construction process, but effectively used seawater can improve the strength development of concretes with slow initial strength development by the inclusion of various admixtures. This can accelerate the demolding time of precast products and contribute to better water-tightness of structures when this is required. Further, reinforcing steel corrosion in concrete using seawater can be inhibited by the use of a corrosion inhibitor or a corrosion-resistant reinforcement material such as stainless steel or steel coated with epoxy resin. Moreover, non-corroding materials such as FRP or bamboo can be used in the right applications.

From the point of view of manufacturing and casting of concrete, seawater concrete makes more effective use of water resources. It can also be expected to improve concrete quality as well as construction efficiency if attention is given to the mix proportion and materials selection, the curing method, and other relevant factors. Many improvements can be expected, for example, improved strength development (initial strength in particular), improved quality (durability) enhancement, earlier demolding, reduced cracking, and improved construction efficiency. Besides serving as mixing water, seawater can also be used as spraying water for wooden molds, to deal with construction joints and for curing (mat curing, ponding, or spray curing), at least if the concrete is unreinforced.

The 21st century has seen design methods in Japan, Europe, and other countries evolve toward the concept of performance-based design. This design method requires only a review to ensure that the required properties are provided over a defined period of time; it imposes almost no limitations with regard to the actual construction methods and materials used. This means, in the context of seawater concrete, that it is only necessary to show (by a certain investigation, test, etc.) that, for example, cracking will not occur during a specified period. Thus, it is certainly possible to use seawater based on the performance-based design method, using as reference the studies, research, and other information presented in this book.

Index

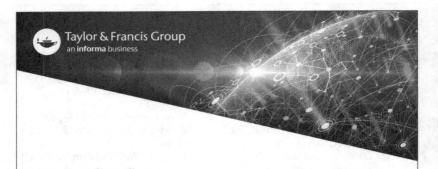

Printed in the United States
by Baker & Taylor Publisher Services